Contemporary Issues in Technology Education

Technology education is a developing field, new issues keep arising and timely, relevant research is continually being conducted. The aim of this series is to draw on the latest research to focus on contemporary issues, create debate and push the boundaries in order to expand the field of technology education and explore new paradigms. Maybe more than any other subject, technology education has strong links with other learning areas, including the humanities and the sciences, and exploring these boundaries and the gaps between them will be a focus of this series. Much of the literature from other disciplines has applicability to technology education, and harnessing this diversity of research and ideas with a focus on technology will strengthen the field.

Occasional volumes on a bi-annual basis will be published under the Council for Technology and Engineering Teacher Education (CTETE) inside this series.

For more information, or to submit a proposal, please email Grace Ma: grace.ma@springer.com

More information about this series at https://link.springer.com/bookseries/13336

Jonas Hallström · P. John Williams
Editors

Teaching and Learning about Technological Systems

Philosophical, Curriculum and Classroom Perspectives

Springer

Editors
Jonas Hallström
Linköping University
Norrköping, Sweden

P. John Williams
Curtin University
Perth, WA, Australia

ISSN 2510-0327 ISSN 2510-0335 (electronic)
Contemporary Issues in Technology Education
ISBN 978-981-16-7718-2 ISBN 978-981-16-7719-9 (eBook)
https://doi.org/10.1007/978-981-16-7719-9

This Springer imprint is published by the registered company Springer Nature Singapore Pte Ltd.
The registered company address is: 152 Beach Road, #21-01/04 Gateway East, Singapore 189721, Singapore

This book is dedicated to the memory of our dear friend and colleague Prof. Jacques Ginestié. Jacques was a professor at Aix-Marseille Université, France's largest university, and the director of the Higher School of Teaching and Education (ÉSPÉ) in charge of teachers' education for primary and secondary school. Despite his gentle nature and softly spoken ways, he was a vociferous advocate for technology education, particularly in France, but also in many other countries. He was the holder of UNESCO Chair in Scientific and Technological Education and Teacher Training, a member of the national monitoring committee for teacher training and the president of the scientific council of the African Network of Training Institutions for Technological Education Trainers (RAIFET). We fondly remember his profound humanism, generosity of spirit and love of life and seek to emulate his professionalism and scholarship.

Foreword

I am delighted that Springer has commissioned a book that brings so many experts together to address the various aspects concerned with the teaching and learning of technological systems. It is a key area of technological knowledge that so many teachers in different countries must address. In New Zealand for example, right from level 1 (6–8-year-olds) students need to 'understand that technological systems have inputs, controlled transformations and outputs', and the 2020 Standards for Technological and Engineering Literacy in the USA put 'systems thinking' as the first of what they consider to be the eight technology and engineering practices. The design and technology curriculum in England for 11–14 year olds states that these students should 'understand' how mechanical systems and electrical and electronic systems can be used in the products they make. I confess that as a schoolteacher, 'systems and control' were the areas of the technology curriculum that I found were the most difficult to teach. Why was that?

First, the idea that everyday items such as an electric oven, the central heating in my home, my mobile phone and the simple streetlamp outside my house—that comes on when it is dark but goes off again from 2 am to 4 am to save electricity—can be usefully modelled in their action by thinking of them as systems are not intuitively obvious. After all, for centuries, medicine was taught in a holistic way as a balance of 'humours' before ideas of considering the human body as a series of biological systems were considered. Organs working together to perform a certain task such as the digestive system, blood circulation system and the nervous system are now considered as a functional block that does a job—but with component parts. For the blood example, components are the heart, blood and blood vessels and for the nervous system the brain, spinal cord and peripheral nerves. In technology, system and control ideas are often used in two different ways, for designing new systems and for analysing existing ones. In much of the teaching I did in school, we used the ideas to *describe* the many systems and processes which we find around us. We use a DVD player, computer and television from a very young age without much idea of what goes on inside the case. Introducing ideas of thinking about and analysing such devices as a system that we can control is a whole new way of thinking.

Second, there is the use of a new 'language' of systems and control. There are many topics we teach where the vocabulary we use can be confusing. For instance, in everyday life we talk about the Police Force, Air Force, energy drinks and energy cereal bars, going out to work and working hard when we study; but 'Force', 'Energy' and 'Work', of course, have very specific meanings in science which students need to learn and use correctly. Learning new and specific vocabulary seems to me to be a particularly acute problem when addressing concepts in systems and control: goal, primary input, secondary input, open-loop control, closed-loop control, comparator, feedback, lag and stability—all have specific meanings which students need to understand.

Third is specifying the level of focus when considering a system. I suppose I mean that I found it difficult when asking students to consider different technological systems to clearly specify to them where the system boundary should be. For example, if they are asked to analyse a mobile (cell) phone as a system that they control, what is it reasonable to include? Yes, they can focus on what might be inside the case, but a mobile only works as a phone by the consideration of the wider cell system of phone masts which pick up and amplify the phone's digital signal and base stations that route on the call. I also found the consideration of the system boundary particularly difficult to teach when I moved from *analysing* existing systems to *designing* new ones. If the design problem was relatively simple, such as a light sensor, the *input* stage would be a 'light-sensing unit' incorporating a light-dependent resistor, the *output* stage an indicator such as an LED or buzzer unit, and the *processing* stage would be a circuit that can make the circuit sensitive to small changes in light level—a 'comparator' board. All well and good. But designing an *input–process–output* system for a wider real-world context can focus on the artefact being designed and ignore the wider social factors that might better address the intended goal.

A few years ago, for example, I was talking about an examination entry by a sixteen-year-old pupil with his teacher. The pupil had designed and made a 'panic alarm' in case he was attacked late at night. In a technical sense, the system was very well done indeed with proper consideration of the alarm's weight, power supply, loudness, ease of action and so on. If anyone had attacked that boy, everyone would have heard about it! I asked his teacher, who was very proud about what his student had achieved, whether the student had considered the issue of *why* such an alarm was needed in his neighbourhood. The teacher looked puzzled by the question as he obviously thought it irrelevant; why such a panic alarm was needed (in terms of the wider values exhibited by those in the pupil's locality) was not part of the examination marking scheme and so not important to consider as it did not gain marks. However, I wondered if by drawing the system boundary narrowly around the alarm itself, this was the best solution to the problem he faced. By not considering *why* he was afraid at night due to few late-night buses or limited and poor street lighting, his solution was, in some senses, restricted. The alarm worked but did it address his fear of going out late in his locality? Maybe the sixteen year old could not do much to himself about the wider context of supplying maybe free buses or better street lighting. However, the narrow system boundary around the well-crafted and technically sound panic alarm provided only a partial solution to the youth's problem, nor a consideration of alternative approaches to crime prevention. A wider system view could consider not just burglar and panic alarms looking at the result of crime, but also engage with the possibility of changing the behaviour of the thieves—soft system thinking as well as hard system thinking.

Looking at the sections of this book, I am sure that I will be able to tackle the teaching about systems and control with increased confidence. Moving from the consideration of the background in thinking and modelling technological systems, to the active consideration of different systems and then on to the discussion of learning and teaching about technological systems is just what I need to address the three aspects of the topic that I find so difficult to teach. I am sure that, like me, you and your teaching will similarly benefit from this excellent book.

Frank Banks
Open University
Milton Keynes, UK

Frank Banks is Emeritus Professor of Teacher Education at The Open University in the UK where he directed the innovative online initial teacher education programme. Frank has worked as a schoolteacher of science, technology, engineering and mathematics in different high schools in England and in Wales.

Introduction

The complexity of technological systems has progressively developed throughout the history of socio-technological revolutions. The currently dubbed Fourth Industrial Revolution (IR4.0) is characterized by cyber-physical systems in which physical and software components are deeply intertwined, able to operate on different spatial and temporal scales, exhibit multiple and distinct behavioural modalities and interact with each other in ways that change with context (US National Science Foundation), all linked through the 'Internet of Things'.

Such systems are ubiquitous for all human beings, to the extent that they remain invisible or inscrutable to most people. Since these technological systems are interdisciplinary and consist of many components with complex connections, they are difficult to understand. Consequently, systems must form a part of every child's education as they progress their technological literacy.

The book deals with teaching and learning about technological systems, in technology education and adjacent curriculum areas. The overall aim of the book is to describe, analyse and synthesize contemporary research on technological systems in technology education, as a resource and reference work for professionals in this area.

The book consists of three distinct parts, between which there is a progression. Part I: *Foundations of a Technological Systems Philosophy* presents the reasonably well-researched underpinnings of systems education. Part II: *Contents, Concepts, and Contexts for Teaching About Technological Systems* reviews the less well-researched areas related to curriculum, systems thinking and models of systems. Part III: *Learning and Teaching About Technological Systems* addresses in more detail the learning and teaching of technological systems. There is little research-informed knowledge about this final part, and consequently, much research remains to be done.

Thus, the book both summarizes prior research and contributes new research, intertwined with expert commentary. The final chapter in Part IV is a synthesis of the research presented in the book.

Perth, Australia P. John Williams

Contents

Editors and Contributors

About the Editors

Jonas Hallström is Ph.D. and Professor of Technology Education at the Department of Behavioural Sciences and Learning, Linköping University, Sweden, where he is also currently Deputy Head of Department. He is also Chair of the Swedish National Graduate School in Science and Technology Education Research (FontD). He presents regularly at international conferences and consults on technology education. He is one of the editorial board members of the Brill/Sense book series *International Technology Education Studies*. His research primarily concerns the historical emergence of technology as knowledge content in the school, the epistemology and subject philosophy of technology, various subject content (e.g. technological systems) as well as the attitudes to and knowledge of technology and technology education of students, student teachers and teachers. The research also relates to technology teaching in relation to, e.g. design, gender (girls and technology), authentic learning, models and modelling and STEM (science, technology, engineering, mathematics) education.

P. John Williams is Professor of Education and Director of Graduate Research in the School of Education at Curtin University in Perth, Western Australia, where he teaches and supervises research students in STEM and technology education. Apart from Australia, he has worked and studied in a number of African and Indian Ocean countries and in New Zealand and the USA. His current research interests include STEM, PCK and electronic assessment of performance. He regularly presents at international and national conferences, consults on technology education in a number of countries and is Long-standing Member of eight professional associations. He is editorial board members of the Springer *Contemporary Issues in Technology Education*and is on the editorial board of six professional journals. He has authored or contributed to over 250 publications, and is elected to the International Technology and Engineering Education Association's Academy of Fellows for prominence in the profession.

Contributors

Moshe Barak Ben-Gurion University of the Negev, Beer-Sheva, Israel

Marjolaine Chatoney Aix Marseille University, Marseille, France

Osnat Dagan Beit Berl College, Beit Berl, Israel

Marc J. de Vries Delft University of Technology, Delft, Netherlands

Susanne Engström KTH Royal Institute of Technology, Gothenburg, Sweden

Fabrice Gunther Aix Marseille University, Marseille, France

Jonas Hallström Department of Behavioural Sciences and Learning, Linköping University, Norrköping, Sweden

Arne Kaijser Department of Philosophy and History, KTH Royal Institute of Technology, Stockholm, Sweden

Claes Klasander Department of Behavioural Sciences and Learning, Linköping University, Norrköping, Sweden

David Mioduser School of Education, Tel Aviv University, Tel Aviv-Yafo, Israel

Per Norström KTH Royal Institute of Technology, Stockholm, Sweden

Maria Svensson Department of Pedagogical, Curricular and Professional Studies, University of Gothenburg, Gothenburg, Sweden

Ann Zetterqvist Department of Pedagogical, Curricular and Professional Studies, University of Gothenburg, Gothenburg, Sweden

Part I
Foundations of a Technological Systems Philosophy

Chapter 1
Socially Constructed and Society Shaping: Investigating Characteristics of Technological Systems for Technology Education

Jonas Hallström and Arne Kaijser

Abstract It is important for students to have knowledge about the basic characteristics of technological systems, because they differ in crucial respects compared to single technological artefacts. Moreover, many technological systems have a far-reaching impact on society and the environment, while at the same time they are socially shaped and controlled by human beings. In the literature such systems are often called *socio-technical systems*. The aim of this chapter is to investigate the key characteristics of technological systems in time and space, in order to inform teaching of technological systems. The specific characteristics of technological systems include them being: socio-technical with both societal and technical components; developed by system builders and managed by professional organizations; with a spatial scope ranging from local/city-wide, regional, national to global networks; dependent on control features including feedback loops as crucial mechanisms for making the systems stable. These technological systems also evolve—and sometimes devolve—in distinct phases and in particular societal, economic and geographical contexts, which may have repercussions when they are transferred. Furthermore, the systems are dependent on each other which over time and space leads to an entanglement of systems. Technological systems, finally, have a huge impact on the environment, which is why students will need to critically consider the human dependence on systems. This chapter introduces a number of systems concepts that might also be fruitfully used as educational concepts in teaching about technological systems. This way students can learn to generalize knowledge of technological systems, so that they can take on, understand and critique different kinds of systems, even ones that have not been designed yet.

Keywords Technological system · Historical development · Geography · Technology education · Control theory · Momentum

J. Hallström (✉)
Department of Behavioural Sciences and Learning, Linköping University, S-60174, Norrköping, Sweden
e-mail: jonas.hallstrom@liu.se

A. Kaijser
Department of Philosophy and History, KTH Royal Institute of Technology, Stockholm, Sweden
e-mail: arne.kaijser@abe.kth.se

J. Hallström and P. J. Williams (eds.), *Teaching and Learning about Technological Systems*, Contemporary Issues in Technology Education,
https://doi.org/10.1007/978-981-16-7719-9_1

1.1 Introduction

> Technological systems contain messy, complex, problem-solving components. They are both socially constructed and society shaping (Hughes, 2012 , p. 45)

This chapter considers the challenge of dealing with technological systems in technology education. It is important for students to have knowledge about the basic characteristics of technological systems because they differ in crucial respects compared to single technological artefacts. Moreover, many technological systems have a huge impact on society and the environment, and still it is crucial to realize that they are also socially shaped and controlled by human beings. In the literature such systems are often called *socio-technical systems* to underline that they consist not only of technical components but also of the people and organizations that design, build and operate these components, as well as the legal and economic frameworks for these activities (Hughes, 1983).

Examples of such technological systems are systems with a global reach such as the Internet or the civil aviation system, as well as national or local systems such as railroads, tramways, water and sewerage or district heating systems. A common characteristic of these systems is that they facilitate movements of different kinds; of people, goods, energy and/or information. Moreover, all these systems, regardless of their size, consist of many components with complex connections and—often—a fuzzy system border (Hallström & Klasander, 2020; Ingelstam, 2012; Joerges, 1996). In addition, they all, to some degree, include control features such as feedback mechanisms (Hughes, 2004). These systems are thus generally more difficult to understand than artefacts, which is a potential challenge from a democratic and technological literacy point of view since all technology needs to be critically evaluated. Technology education in schools therefore needs to address such complex technological systems, and how to manage and critique them in an informed manner (e.g. Keirl, 2006, pp. 97–99).

What makes the task of understanding these systems and how they have evolved over time so important is their enormous impact on society. In the past two centuries, the introduction and expansion of large technological systems have led to fundamental societal changes. On a macro level, these systems have spurred a transition from a basically "nature-based" economy, where the location of industries and other economic activities was primarily dependent on the access to waterways and natural resources, to a "system-based" economy, where easy access to technological systems has become decisive for the location of most economic activities (Högselius et al., 2016; Kaijser, 2003). The systems have enabled an intensified exploitation of natural resources as well as a division of labour on a hitherto unknown scale. Moreover, technological systems have contributed to rapid urbanization. In short, the development of technological systems is an important factor in the sustained economic growth in the past two centuries. On a micro level, technological systems have resulted in what could be called a "convenience revolution" for everyday life. Many of the most strenuous household tasks have been taken over by electric household appliances, tap water and central heating. Furthermore, the car and the telephone have given many households a dramatic increase of mobility and reach.

In this chapter, we will be following the historical evolution of technological systems because this is a fruitful way of disentangling the complex characteristics of the systems that have accreted over time (Arbesman, 2017). Similarly, tracing the spatial networks of pipes, cables, roads, base stations, etc. of such systems may also make them more transparent. The aim of this chapter is thus to investigate the key characteristics of technological systems in time and space, in order to inform teaching of technological systems. To that end, the chapter presents how such systems have developed over time and space in relation to aspects of control and control thinking. Hereafter, the term technological system will be employed but several examples and concepts used in the chapter relate to Thomas P. Hughes' conception of socio-technical systems (Hughes, 1983, 2012). These and other concepts and examples are drawn from a selection of literature primarily in the philosophy, sociology and history of technology, but also in control theory and complexity science.

1.2 Definition and Characteristics of Technological Systems and System Designers/Builders

A very basic definition of a technological system is that it fulfils a particular purpose, consists of components with relations, connections, or flows between them, and that there is a system boundary. Beyond the system boundary there is the surrounding, which may interact with the system but is not part of it (Ingelstam, 2012). Technological systems can be more or less open to exchange of energy, matter and/or information with the surrounding (Hallström & Klasander, 2020). Furthermore, they can to a lesser or greater degree be considered socio-technical systems, that is, including both technical and societal components. According to Hughes (2012), large technological systems include not only technological components—the "technical core"—but societal aspects, too, including engineering companies, banks, legislation and research. People, for example the engineers who design and build the system, are also considered part of the system, in his view.

Technological systems have become significant from a democratic point of view in many countries by solving problems and offering services to increasingly larger sections of the population (Kaijser, 2003; Melosi, 2008). Many such systems have become a vital part of society's infrastructure, and have taken over many of the tasks that were previously carried out by people in homes, workplaces and public spaces, such as collecting water, managing heating, sharing news, transporting people, information, providing energy sources, etc. Some researchers denote these as *infrastructural systems* or just infrasystems (Högselius et al, 2016; Kaijser, 1994). However, the access to the systems and their services is not equally distributed, which leads to issues of social injustice (Kates et al., 2006).

A major challenge of large technological systems is that they can have detrimental effects on the environment. The growing number of cars, airplanes, fossil power plants, etc. is responsible for a large part of all greenhouse emissions. Moreover, enormous amounts of resources are needed for the construction and use of these systems. In this sense they can be seen as global *socio-metabolic* systems, needing resources that transcend the local place of origin. It is obvious that fossil-based systems have huge "ecological footprints" at oil and gas fields and coal mines far away from the user. Less obvious is that also solar-power or wind-power plants require vital resources from far away, not least rare earth metals. Thus many technological systems have become a socio-metabolic system on a global scale, with an "ecological footprint" much larger than the land and geographical space for which it was designed (cf. Hornborg, 2020; Rees, 1997).

One of the earliest proponents of modern, urban technological systems was the Englishman Edwin Chadwick. Originally a barrister, Chadwick developed an interest in sanitary reforms amid the dirty, unhealthy urban environments of 1830s and 1840s Britain. His was a technological solution: technological water and sewerage systems would improve and cleanse the environment and prevent disease. His vision borrowed from nature, and he compared water supply and sewerage systems with the artery and vein systems of the human body. Contemporary English engineers did not adopt Chadwick's grandiose systems ideas immediately, but Chadwick had a major influence on views about water and sewerage technology in Britain, the rest of Europe and the USA during the latter half of the nineteenth century. Water and sewerage were increasingly seen as integrated technological systems (Hamlin, 1992; Melosi, 2008).

Chadwick can be regarded as a *system builder*, a term used by Hughes (1983) to describe those who initiate, build, control and/or expand large technological systems, regardless of whether they are inventors, designers, engineers, scientists, administrators or investors. Another important system builder of the nineteenth century was Thomas Edison, who used the existing gas systems as a model when he built the first electric lighting system in the banking district of Manhattan. He further developed electricity systems to form large, centralized grids. These and many other system builders were visionaries who understood and took advantage of the systemic properties of technology (Bladh, 2006; Hughes, 1989), which then proliferated during the twenty and early twenty-first centuries with a number of types of technological systems. The notion of connecting together technological artefacts into networks or systems that fulfil various functions and are humanly controlled, proved resilient and successful.

1.3 Technological Systems as Control Systems

A salient feature of technological systems is the fact that they impose control on humans, society and the environment. Sophisticated drainage and sewerage systems have been used for millennia to help control rivers and establish human settlements, for instance (e.g. Kaijser, 2002; Mays et al., 2007; Nia et al., 2019). In the early to mid-nineteenth century, technological systems were introduced in a variety of urban settings in order to impose control of water provision, waste management, energy provision and more. In this way, Chadwick, Edison and other system builders managed freshwater, waste and energy management and lighting, etc. in the increasingly controlled urban environment. As these systems gradually became an indispensable part of daily life, breakdowns in their operations caused much inconvenience. Their reliability and robustness therefore became of increasing importance (Högselius et al., 2013).

Operating these increasingly complex systems subsequently became a pivotal issue in the twentieth century, which spurred the development of modern control engineering, in which the feedback mechanism is essential. James Watt's centrifugal flyball governor in his late eighteenth century modified steam engine was the first simple yet effective feedback mechanism to be widely used. More complex uses of feedback control were devised in the early twentieth century for modern systems. Examples of such systems prior to the Second World War include robust amplifiers for long-distance telephony, and the use of analogue computers for designing resilient regional electricity systems (Mindell, 2002). The main figure in modern control engineering is the mathematician Norbert Wiener. He was involved in efforts during the Second World War to design anti-aircraft gunfire-control devices, and this experience inspired him to develop cybernetic theory of human–machine and machine–machine interaction after the war, which became the foundation of modern control theory, pivotal in all subsequent technological systems (Wiener, 1948, 1950/1989).

The essential feature of operating modern complex technological systems is thus by way of *feedback*, especially negative feedback for making systems operate in a stable and robust way, like in James Watt's flyball-regulated steam engines. Modern control systems use sensors to communicate information back to a regulator unit in order to control the system, thereby aiding operators or making entire processes automatic (Glad & Ljung, 2000; Mindell, 2002). Feedback has many advantages, the most important being that it makes the operation of a complex system more robust against disturbances of various kinds. Thereby it increases the reliability of the system.

However, the control of systems is not only about avoiding technical breakdowns. Another important issue is to use the capacity of a system in an optimal way; most importantly, to avoid surpassing the capacity leading to breakdowns, but also to avoid underutilizing the system. In the late nineteenth century, electrical engineers in the USA introduced the concept of *load factor*, defined as the average load divided by the peak load (Hughes, 1983). By using smart tariff systems, power companies could give incentives for example for large industrial customers to use most of their electricity at night or in the middle of the day, when household use was small. Similar principles of capacity pricing have later been adopted in public transport, aviation, telephony, etc., mostly without much controversy. However, efforts to introduce road tolls for reducing rush hour traffic in large cities have been controversial and opposed by the motorcar lobby (Isaksson, 2008).

1.4 Technological Systems Over Time

Technological systems undergo fundamental changes over time. One way of describing these changes is by distinguishing different developmental phases: establishment, expansion and stagnation. This does of course not imply that all systems go through exactly the same, predetermined development. However, the problems and challenges that system builders and other actors confront differ considerably over time as the systems evolve, and thus these problems/challenges define the three phases.

1.4.1 Establishment

Many technological systems are based on what Hughes (2012) calls *radical inventions*, "because they inaugurate a new system" (p. 51). Radical inventions are connected with inventors such as Alexander Bell, Thomas Edison, Guglielmo Marconi and many others. To establish such a radical invention to initially form a system generally requires a huge financial investment, and at this early stage it is mostly very difficult to assess the future demand for the services of the system. The establishment phase is therefore characterized by a very high *uncertainty*, and a crucial problem is how to design mechanisms for overcoming this uncertainty. Here it is important to emphasize that many attempts in the past at establishing new technological systems did not succeed. The actors behind them were not able to overcome the initial uncertainty and create credibility for their plans.

A common feature of many successful technological systems is therefore that a radical technical invention was also followed by an *institutional innovation* made at an early stage, enabling a common use of the new system by many different groups, thereby sharing investment costs and diminishing uncertainty. The introduction of gas lighting illustrates the importance of coupling a new technological system with such an institutional innovation. In the late eighteenth century, many inventors tried to develop new technologies for lighting that were cheaper than the dominant light sources at this time: tallow and whale oil. An Englishman, William Murdoch, designed a simple gasworks in which gas could be produced out of coal, peat and wood. In 1802, Murdoch installed gas lighting in the premises of the Boulton & Watt Company in Birmingham, and it radically reduced the lighting costs of the company. In the following decade some dozens of gas systems were sold to other factories. However, a gasworks represented a considerable investment in retorts and pipes, and thus the market for the new lighting system seemed to be restricted to large factories with a high demand for lighting (Kaijser, 1986).

Around 1810, a new idea was developed for how to overcome this obstacle by the German entrepreneur Friedrich Albert Winzer. His idea was to sell gas, not gasworks. He realized that a precondition for gas lighting to reach a big market was that the investment cost for a gasworks could be shared by many users. He developed a plan to establish a joint-stock company that would build a big gas-producing plant in the middle of London, and a whole network of pipelines under the streets. This would enable the company to sell gas at a relatively low price to a large number of subscribers. He was successful in convincing actors with financial and political weight to support his plan, and in 1814 the "The Gas Light and Coke Company" in London began its gas sales. Within ten years, gas lighting was used by many thousands of subscribers in shops, restaurants, workshops, offices and households as well as for street lighting. Many other cities in Britain and on the continent followed London's example in the next decades (Kaijser, 2003).

The idea of selling gas instead of gasworks became an institutional innovation of fundamental importance. It was by finding a way for a communal use of the expensive gas-producing plants that gas lighting became affordable for many more. A similar story of a common use of a system by many different kinds of social groups can be told for a number of other technological systems. In the railway system, a key institutional innovation was to provide not just a rail but to offer transport. The first public railway, built in 1825 between Stockton and Darlington in England, was organized like a canal, providing a rail which customers with adequate wagons and horses or locomotives could use. This proved to be inefficient when traffic grew, and the second public railway built in 1830, between Manchester and Liverpool, was therefore organized in a very different way. In addition to building a railway, this company also bought wagons and locomotives and organized train traffic following elaborate timetables. Likewise, for achieving a fast expansion of the postal and telegraph systems, it was essential that many kinds of interests (civil service, armed forces, railway companies, business persons, newspapers, etc.) could share the same service (Kaijser, 2003). In the case of urban water systems, a prerequisite for mustering the necessary capital to build waterworks, water towers and pipes was that the water was seen as a common good that could be used for several purposes; in households, in factories and for firefighting (Hallström, 2002).

When one city or region had been able to establish a successful technological system, many others would often soon want to follow its example. However, each city, region or country had to adapt the institutional set-up of the system to its own political and socio-economic conditions. It should be noted that some technological systems were easier to establish than others, depending on the character of their network. The above examples are all systems that needed a *specific* technical network of their own—gas pipes, rails, telegraph lines, water pipes—and which, in addition, were very expensive to build. Other systems were able to use *existing* transport or communication networks. The postal system is a clear example since it utilizes existing networks for transportation (roads, rails, etc.). The Internet could also at an early stage use existing telephone networks. A third category of systems has been able to use *nature given* networks like water, air or the electro-magnetic ether in conjunction with harbours, airports, transmitters and receivers. The latter two kinds of networks need much less investments at an early stage (Hallström, 2009; Kaijser, 2004).

1.4.2 Expansion

Once a technological system has been established and reached a first major market an entirely new situation develops. The revenues from sales provide an economic base and the experiences of building and operating the system often lead to further technical improvements. This shapes the prerequisites for the expansion of the system, and there are generally strong economic and social forces for expansion. When trying to expand the technological systems, system builders often face technical obstacles and difficulties threatening to block the expansion.

Let us first look at the economic forces. The marginal costs for providing additional units of service have usually decreased in expanding systems, due to *economies of scale* and *economies of scope*. Economies of scale arise primarily on the "production side". For example, the production cost for a unit of gas or electricity decreased when the size of the generating plants increased. Likewise, the cost per passenger or unit of goods usually decreased as the size of ships, trains and airplanes increased. Falling costs enabled lower prices, which raised demand and spurred further increase of scale, etc. Economies of scope arise primarily on the consumption side. For example, gas was first used mainly for gas lighting, which implied that most of the gas was used in the morning and in the evening. In the late nineteenth century, gas was also introduced for cooking, water heating and for industrial processes. These uses had different demand patterns over time, resulting in a more even use of gas, a higher *load factor* of the system. A high load factor meant a better and more even technical utilization of the system and thereby also a better utilization of the huge capital costs of the system, which led to lower costs per unit of service. Similar increases of the load factor have taken place in many other systems as well, when new categories of customers or new types of services have been introduced. This has often been achieved through very deliberate policies from the system operators offering the new users favourable prices. Thus, while economies of scale arose through an increase of the size of the technical components of a system, economies of scope arose through more diverse uses of the system (Kaijser, 2003).

In technological systems providing communicative services, there is also another kind of economic force stimulating expansion that we call *economies of reach*. The economies of reach have to do with the extent of the network, and thereby the number of people or places that can be reached by using it. In a telephone system, for example, the connection of a new subscriber is not only rewarding to the new subscriber, but also to all the existing subscribers who receive an additional person to call. At the turn of the century 1900, there were two competing telephone companies in Stockholm, and they tried to increase the attractiveness of their networks by offering telephone subscriptions for free to doctors, pharmacists and other professionals that their subscribers wanted to reach. Economies of reach have also been of importance for the growth of transport systems. For example, the attractiveness of having a car increased as the network of roads grew and was improved. In many countries, powerful motorcar lobby organizations arose which succeeded to persuade authorities to invest in the improvements of roads. Such improvements literally paved the way for a fast increase in car ownership, which in turn formed an economic base (through taxes on cars and on fuel) for further investments in the expansion of roads (Hallström, 2009; Kaijser, 2003).

Economies of reach can also be said to apply to the Internet. In the USA, the National Science Foundation long administered the Internet but handed it over to commercial interests in 1995. In Europe in 2001, there were no less than 27 privately owned Internet Service Providers (ISPs), most of them offshoots from universities, operating a pan-European network (Högselius et al., 2016, pp. 58–59). In the early 2000s, the Internet expanded rapidly, especially after 2007 with the introduction of the smartphone (iPhone). Over time, this introduction greatly expanded the extension of the Internet due to the possibility of using it via the smartphone, and consequently the smartphone led to a strong economies of reach effect. The new smartphone user got access to the whole Internet and its millions of users at the same time as a new user was added (e.g. Briggs & Burke, 2009).

Besides the economic forces there have also been strong "social" forces for expansion. Expanding technological systems acquire what Thomas P. Hughes calls *momentum* (Hughes, 2012). Companies which invest much capital and other resources in a system–both equipment producers, owners and operators of systems–develop a strong interest in the further expansion of the system. Furthermore, most systems need engineers with special skills, and these often have a common educational background. Such professional groups form tightly knit networks sharing the same values and with similar views concerning the desired future direction of the system. They thus develop a common *system culture,* to use another concept introduced by Hughes, which meant that close co-operation among many influential actors towards a common goal became a strong force for expansion of a system (Fridlund, 1999; Hallström, 2002).

Thus, once a technological system has been established on a first major market, strong forces for expansion arise—economies of scale, scope and reach—producing a spiral of growth. However, a fast expansion of a system is seldom a smooth process. Technical obstacles and difficulties often appear, threatening to block the expansion. Thomas P. Hughes uses a military metaphor to describe this phenomenon. He talks about a *reverse salient* in an advancing front, which was a typical feature in the trench warfare during W.W.I. When such reverse salients emerged in the expansion of systems, no resources were spared to try to overcome them. The best scientists and engineers available were engaged in efforts to develop new components, sub-systems or system architectures. When these were implemented, they enabled a continued fast growth (Fridlund, 1999; Hughes, 2012).

A fast expansion of a technological system, brought about by alliances of powerful actors, often resulted in considerable economic gains for many parties. However, in many cases the system also led to negative effects in the form of pollution and health problems affecting other, less influential people. An example of this is the introduction of water closets in the apartments of well-to-do urban households at the turn of the century 1900. Water closets improved the hygienic conditions for their users, but they also contributed to the pollution of rivers and lakes and, in turn, the freshwater sources of many other people further downstream (Hallström, 2009). A similar example is the building of city airports in the mid-twentieth century, enabling fast transport for businessmen and rich households but causing considerable noise pollution as well. In the past, such environmental and health problems have generally not become reverse salients in Hughes' sense, because they have not affected influential actors. Therefore, they have not attracted the same kind of attention as the direct technical, economic and institutional obstacles to growth.

The tendency of technological systems to sometimes limit actors' scope for action and cause momentum is also clear when looking at the changed *functions* of systems. When it comes to sewerage, systems were designed and built in a piecemeal fashion until the early to mid-nineteenth century, and their purpose was up to then simply to drain marshland and lead away rainwater where necessary. With the introduction of modern, urban systems in the late nineteenth century, however, the function shifted towards becoming a large waste management system. The wastewater system subsequently had to deal with more types of waste, some of which were detrimental to the system and had to be banned. Yet both wastewater engineers and users became accustomed to a certain way of using the system, making it harder to change but also—once the initial resistance to the water closet had been overcome—easier to continue to expand. Sewage treatment plants improved the sewerage system from a health and environmental point of view, which made sewerage even more firmly established in the mid- to late-twentieth century (Drangert & Hallström, 2002; Hamlin, 1992; Melosi, 2008).

There was a similar functional development in the growth of the Internet. It also started out with piecemeal extensions and limited use until the 1990s, when the Internet really took off as a new societal phenomenon labelled variously as an information or digital highway. The key to the success of the Internet was the *convergence* of an increasing number of media and communications features into the same system: e-mail and chat groups; web pages, blogs/vlogs, online forums; social media; media streaming and video calls; e-commerce, online tutoring, etc. (Briggs & Burke, 2009). As with sewer systems, with each new innovative function the momentum of the Internet grew, but on a much larger, global scale (Hughes, 2012). Each new additional feature added to an increased functional complexity of the Internet, not the least the myriad of different control sub-systems and feedback loops required to keep the system stable. There was simultaneously also an *accretion* of technical components where new ones were built upon old ones, making the system less and less perspicuous (Arbesman, 2017).

1.4.3 Stagnation

Expansion processes do not go on forever. Eventually technological systems reach a stage when growth rates diminish and a phase of stagnation commences. One factor contributing to stagnation is a weakening of economic forces for expansion. Economies of scale have reached a saturation level in a number of systems. For example, the maximum size of power plants, freight ships and airplanes has not increased since the 1970s. Likewise, economies of scope are usually more or less exhausted once the load factor of a system has reached a certain level. When it comes to economies of reach, some systems have even experienced a transformation into *dis*economies of reach. In the case of road traffic in urban regions, additional cars have steadily increased congestion. The more cars, the longer it takes for each person to reach his or her destination, which hampers the further expansion of car traffic. Moreover, many systems have experienced a saturation of the demand for their services, partly due to more efficient end-use technologies, such as LED lamps or washing machines using much less water and electricity than before.

There is one more factor which has often played a crucial role in processes of stagnation and decline of technological systems, namely competition from other systems providing similar services. During most of the nineteenth century, gas systems were uncontested as providers of high-quality energy that could be used for lighting, cooking, heating and even mechanical power. However, the electric power systems established in the early 1880s provided an energy source which could be used for exactly the same purposes. Consequently, a fierce competition arose between the promoters of the two systems, stimulating technical improvements in both systems. Around 1900 the struggle for the lighting market therefore led to a dramatic increase in the efficiency of both gas lamps and electric lamps, but in the 1910s electric lighting had become superior and gas lighting declined. In many countries, gas systems were pushed back also in the other markets and many gasworks were closed down in the mid-twentieth century (Kaijser, 2003).

Similar processes of competition have taken place among transportation and communication systems. In the second half of the nineteenth century, canals competed with railways (e.g. Torbrand, 1978) and half a century later there was an intense struggle between railways and motorcars. Likewise, the telegraph and the postal systems had to struggle with the telephone system in the beginning of the twentieth century. The older systems struggled very hard to improve their efficiency, and while being pushed back in some markets they were sometimes able to keep their position in some market segments in which they had special competitive advantages. Still, the earlier expansion was replaced by stagnation.

1.5 Technological Systems in Space

A technological system is always designed for a particular time and place, and is therefore technologically, administratively and politically closely linked to a historical and geographical context. The spatial scope varies significantly between different technological systems. Gasworks, water and sewage systems, tramways, electricity and telephony are examples of infrastructural systems originally intended for urban areas, providing households and businesses with specific services by way of specially constructed networks of pipes, lines or rails. The investment cost in these networks was high, but the density of cities made it possible to reach many potential customers and cover these costs. Other systems, like canals, railways and telegraphs, were intended for facilitating transport or communication between cities or between factories or mines and a sea port, usually within a country. In these cases, a high demand of transport or communication between distant places was a prerequisite for the high investment in the network. In addition, the postal system was created for communication between far away cities and towns, and in this case existing transport networks could be used. When civil aviation systems were established in the inter-war period, the main purpose was to facilitate passenger transport across national borders (in Europe) or state borders (in the USA) (Kaijser, 2004; Vermaas et al., 2011).

Already in the establishment phase, technological systems can thus have very different spatial extensions. Furthermore, the expansion of these systems can take different forms. One important kind of expansion is through transfer of a certain system to other places. The first gasworks in London became a model for many other cities, first in Britain and later in Europe and the USA. The transfer of a system from one place to another is complicated. Or in the words of Thomas P. Hughes: "Because a system usually has embodied in it characteristics suiting it for survival in a particular time and place, manifold difficulties often arise in transfer at another time or to a different environment" (Hughes, 2012, p. 60). Hughes uses the biological term *adaptation*, to describe this process as every technological system must be adapted to a new environment during a transition, in which legislation, organizational culture and social, political and geographic conditions are completely different to those in the original environment (cf. Arthur, 2009). This means that, following adaptation to new conditions, every technological system gets a new *technological style* (Hughes, 2012). For example, Swedish water and sewerage systems were inspired by British predecessors. However, they were adapted to suit Swedish conditions, and a unique Swedish technological style arose. In contrast to British waterworks, which mainly used steam engines to pump water up from rivers, Swedish ones mostly used turbines as hydropower was easily available. In addition, the heavy frost in the ground during the Swedish winter meant that all pipes had to be laid considerably deeper (Hallström, 2002; Isgård, 1998).

Another kind of expansion was by increasing the use of a system, either by stimulating existing customers to increase their consumption (e.g. when households started to use gas for cooking besides for lighting) or by recruiting new customers. This could be done by expanding the network beyond the city borders, to suburbs or smaller towns in the vicinity of a city, or by interconnecting several cities into regional or even national systems. Telephony and electricity systems went through such developments in the late nineteenth and early twentieth century (Kaijser, 1986). This kind of interconnection of local systems gave an incentive for technical standardization; as long as cities built their own systems they could choose the technical design they preferred, but to interconnect they had to adjust to each other, which could be done by defining a common standard for crucial system characteristics such as voltage (cf. Nye, 2006).

It should be emphasized, however, that these kinds of spatial expansions of systems that started in the mid-nineteenth century were often slow and uneven. For example, the rich and influential people in the urban centres were reluctant to extend water and sewerage systems to the suburbs inhabited by poor households or to surrounding rural areas, as this would increase the cost for the provision of these services (Hallström, 2011). There was thus a growing gap in the early twentieth century between urban and rural areas in terms of provision of system services, which partly still remains. In particular, in Africa, Asia and Latin America many households both in the countryside and in urban slum areas still lack running water, sewerage and electricity. Thus, technological systems have often reinforced and exacerbated social differences rather than evening them out (Akallah & Hård, 2020).

The interconnection of technological systems into regional, national and even transnational systems implied thorough organizational and institutional changes, with a concentration of the power over the systems. In most European countries, public authorities took a leading role in the process of system expansion and interconnection, while private companies played a dominant role in the USA (Högselius et al., 2016; Melosi, 2008). The so-called deregulation that has taken place since the 1980s has meant that private companies have taken over many technological systems also in European countries, transferring part of the power from political to corporate entities. The Internet seemed to represent a different pattern from most previous systems. Due to the initially decentralized network character of the Internet infrastructure as well as its organization, it was long hailed as an implement of decentralized power, increased democratization, and amelioration of social and economic differences in society. However, in recent years many researchers and commentators have questioned this role of the Internet, due to the increased political polarization created by "click-baiting" on, e.g. Facebook and Twitter, and manipulation of social media services by Russian "troll factories" through AI-controlled fake accounts (Feenberg, 2017; Gelin & Pettersson, 2018; Sörlin, 2019; Vaidhyanathan, 2018). Some researchers, like Feenberg (2017), still believe in the democratic value of the Internet: "[I]t is still the case that truly free, reciprocal, bottom-up communication has emancipatory potential and such communication does occur on the Internet" (p. 98). One example is the #MeToo movement (cf. Brooks, 2019).

1.6 Entanglement of Systems

Above we have discussed technological systems as single systems. However, many systems have become more and more entangled with each other. Technological systems often play a complementary role to each other achieving synergistic effects. One example is the building of telegraph lines along railway tracks, which began in the mid-nineteenth century. The telegraph made it possible to communicate between stations and this made it possible to increase train traffic considerably. At the same time, the railway facilitated the building and maintenance of telegraph lines, and also provided a guaranteed market for telegraph communication. Thus, building railway tracks and telegraph lines along each other became a standard procedure all over the world. In the early twentieth century, railway companies in countries with abundant hydropower started to electrify the locomotives and built electric lines along—or rather above—the rails. In these countries, electricity was cheaper than coal, and electricity was also more convenient. For power companies, on the other hand, the railways represented a stable and reliable customer. Thus, there were mutual spatial and economic synergies between these two technological systems.

Another example of system synergism is a co-generation plant, in which the heat losses from thermal electricity generation are used for the heating of many houses via a so-called district heating system. The combined production of electricity and heat is almost twice as efficient as a separate production of each, and this was often a major incentive to build district heating systems. A third example is from the transportation sector. Transportation systems need to co-operate because their networks have different coverage, on land, over water or in the air. Railway stations, harbours and airports have since long been crucial nodes in flows of goods and passengers. However, an obstacle to such co-operation was long the high costs of trans-shipment between different modes of transport. In the 1950s and 60s the container was introduced to facilitate the integration of different transportation systems, and thus to achieve synergistic effects (Kaijser, 2003). Since then a growing portion of global commodity flows are handled by containers and mostly flow smoothly, unless a container ship gets stuck in a crucial canal, like the Ever Given in the Suez Canal in March 2021.

The entanglement of technical systems has increased dramatically in the past half century. Almost all transport and energy systems are operated by the help of sophisticated information and communication systems (such as the Internet), and transport and communication systems need energy systems for their operation. This entanglement is ambiguous from a control and reliability perspective. On the one hand, information systems are designed to control the management and operation of systems and to help avoid disturbances or breakdowns. On the other hand, the entanglement implies that a major breakdown in one of the systems may spill over to other systems. For example, major electricity blackouts or hacker intrusions can endanger many other systems (cf. Arbesman, 2017; Hughes, 2004).

1.7 Concluding Discussion and Educational Implications

Technological systems were designed by humans to solve specific problems in particular times and places. In this chapter we have focused on modern technological systems, that is, systems designed, constructed and used after the late eighteenth century in conjunction with industrialization and urbanization. The specific characteristics of such technological systems include them being: socio-technical with both societal and technical components; developed by system builders and managed by professional organizations; with a spatial scope ranging from local/city-wide, regional, national to global networks; dependent on control features including feedback loops as crucial mechanisms for making the systems stable. These technological systems also evolve—and sometimes devolve—in distinct phases and in particular societal, economic and geographical contexts, which may have repercussions when they are transferred to other contexts. These characteristics are common to most technological systems but, for example, the network character and the reliance on control theory for control, communication and automation in and between systems and humans, are unique systems features that students need to learn specifically for each technological system.

The concepts described in this chapter were designed specifically for a scientific study of technological systems—for understanding, designing, using and critiquing them (e.g. Arbesman, 2017; Briggs & Burke, 2009; Dusek, 2006; Hughes, 2004, 2012; Vermaas et al., 2011; Wiener, 1948) —but they may be equally fruitful as *educational concepts* in the classroom when teaching and learning about technological systems at various levels in education. For example, Hughes' (1983, 2012) concepts are straightforward, intuitive and make sense in everyday situations. The term *system builder* refers to someone who builds systems. *Technology transfer*—a concept used by Hughes (2012) but invented long before him—has to do with the transfer of technology and technological systems, to new social, geographic and institutional contexts. With these and the other concepts presented in this chapter it is possible for students to compare technological systems with each other so as to gain knowledge of similarities and differences between systems in different time periods, systems in different geographical areas or institutional-economic contexts, and systems of different kinds. In this way students learn to generalize knowledge of technological systems, so that they can take on, understand and critique different kinds of systems, even ones that have not been devised yet (Hallström, 2020; Hallström & Klasander, 2020; Williams, 2017).

An innate problematic of technology education and technological literacy is how to help students learn about technology in a world where technological development is very rapid (e.g. Dakers, 2006). According to one of the most cited authorities, the American *Standards for Technological Literacy: Content for the Study of Technology*, technological literacy is "the ability to use, manage, assess, and understand technology" (*Standards for Technological Literacy: Content for the Study of Technology*, 2007, p. 7). This is particularly critical when it comes to technological systems which are in themselves already difficult to comprehend and critique (Keirl, 2006), at the same time as they often develop and get more complex over time. Therefore, the use of scientific concepts for educational purposes may be a way forward. In science communication, visualization and visual literacy research, the concept of *exploranation* has recently been introduced to refer to the combination of scientific exploration and educational explanation in one and the same visualized activity (Ynnerman et al., 2018).

Translated to the technology education context, it means that exploranation can be achieved by employing scientific systems concepts, for instance, those devised by Thomas P. Hughes, Norbert Wiener and others presented in this chapter, as educational concepts in teaching about technological systems. For example, students could be tasked to look at the evolution of the Internet and document the multiple functions that have *converged* (Briggs & Burke, 2009) and/or how new technical components have *accreted* (Arbesman, 2017) over time and made the system increasingly complex. Students could also make use of Hughes' concept of *radical invention* (Hughes, 2012) to explore inventions that have been essential in the establishment of new technological systems in different time periods, and make comparisons between the systems. This way, teaching about technological systems can rest on a scientific foundation at the same time as it also makes students more technologically literate.

References

Akallah, J. A., & Hård, M. (2020). Under the historian's radar: Local water supply practices in Nairobi, 1940–1980. *Water Alternatives, 13*(3), 886–901. Retrieved from www.water-alternatives.org

Arbesman, S. (2017). *Overcomplicated: Technology at the limits of comprehension.* Portfolio.

Arthur, W. B. (2009). *The nature of technology: What it is and how it evolves.* Free Press.

Bladh, M. (2006). Hughes teori om Stora tekniska system. *Historisk Tidskrift, 126*(1), 69–90.

Briggs, A., & Burke, P. (2009). *A social history of the media: From Gutenberg to the internet* (3rd Ed.). Polity Press.

Brooks, M. A. (2019). Book review anti-social media: How Facebook disconnects us and undermines democracy. *Information, Communication & Society, 22*(8), 1181–1183.

Dakers, J. R. (Ed.). (2006). *Defining technological literacy: Towards an epistemological framework* (1st ed.). Palgrave Macmillan.

Drangert, J.-O., & Hallström, J. (2002). Den urbana renhållningen i Stockholm och Norrköping: Från svin till avfallskvarn? *Bebyggelsehistorisk Tidskrift, 44*, 7–24.

Dusek, V. (2006). *Philosophy of technology: An introduction.* Blackwell.

Feenberg, A. (2017). *Technosystem: The social life of reason.* Harvard University Press.

Fridlund, M. (1999). *Den gemensamma utvecklingen. Staten, storföretaget och samarbetet kring den svenska elkrafttekniken.* Brutus Östlings Bokförlag Symposion.

Gelin, M., & Pettersson, K. (2018). *Internet är trasigt. Silicon Valley och demokratins kris.* Natur & Kultur.

Glad, T., & Ljung, L. (2000). *Control theory: Multivariable and nonlinear methods.* Taylor & Francis.

Hallström, J. (2002). *Constructing a pipe-bound city: A history of water supply, sewerage, and excreta removal in Norrköping and Linköping, Sweden, 1860–1910.* Department of Water and Environmental Studies, Linköping University, Linköping.

Hallström, J. (2009). Systemteori och teknik. En introduktion till stora tekniska system. In P. Gyberg & J. Hallström (Eds.), *Världens gång - teknikens utveckling. Om samspelet mellan teknik, människa och samhälle* (pp. 99–117). Studentlitteratur.

Hallström, J. (2011). Urban or suburban water? Working class suburbs, technological systems and environmental justice in Swedish Cities in the Late Nineteenth Century. In G. Massard-Guilbaud & R. Rodger (Eds.), *Environmental and social justice in the city: Historical perspectives* (pp. 133–153). The White Horse Press.

Hallström, J. (2020). Embodying the past, designing the future: Technological determinism reconsidered in technology education. *International Journal of Technology and Design Education*, 1–15.https://doi.org/10.1007/s10798-020-09600-2

Hallström, J., & Klasander, C. (2020). Making the invisible visible: Pedagogies related to teaching and learning about technological systems. In P. J. Williams & D. Barlex (Eds.), *Pedagogy for technology education in secondary schools: Research informed perspectives for classroom teachers* (pp. 65–82). Springer.

Hamlin, C. (1992). Edwin Chadwick and the engineers, 1842–1854: Systems and antisystems in the pipe-and-brick sewers war. *Technology and Culture, 33*(4), 680–709.

Hornborg, A. (2020). Machines as manifestations of global systems: Steps toward a sociometabolic ontology of technology. *Anthropological Theory*, 1–22.https://doi.org/10.1177/1463499620959247

Hughes, T. P. (1983). *Networks of power: Electrification in Western society, 1880–1930*. Johns Hopkins University Press.
Hughes, T. P. (1989). *American genesis: A century of invention and technological enthusiasm 1870–1970*. Viking.
Hughes, T. P. (2004). *Human-built world: How to think about technology and culture*. University of Chicago Press.
Hughes, T. P. (2012). The evolution of large technological systems. In W. E. Bijker, T. P. Hughes, & T. J. Pinch (Eds.), *The social construction of technological systems: New directions in the sociology and history of technology* (Anniversary Edition, pp. 45–76). The MIT Press.
Högselius, P., Hommels, A., Kaijser, A., & van der Vleuten, E. (Eds.). (2013). *The making of Europe's critical infrastructure*. Palgrave MacMillan.
Högselius, P., Kaijser, A., & van der Vleuten, E. (2016). *Europe's infrastructure transition: Economy, war, nature*. Palgrave MacMillan.
Ingelstam, L. (2012). *System - att tänka över samhälle och teknik* (2nd ed.). Energimyndigheten.
Isaksson, K. (Ed.) (2008). *Stockholmsförsöket - en osannolik historia*. Stockholmia förlag.
Isgård, E. (1998). *I vattumannens tecken. Svensk VA-teknik från trärör till kväverening*. Ohlson & Winnfors.
Joerges, B. (1996). Large technical systems and the discourse of complexity. In L. Ingelstam (Ed.), *Complex technical systems* (pp. 55–72). Swedish Council for Planning and Coordination of Research (FRN).
Kaijser, A. (1986). *Stadens ljus. Etableringen av de första svenska gasverken*. Liber förlag.
Kaijser, A. (1987). From local networks to national systems: A comparison of the emergence of electricity and telephony in Sweden. In F. Cardot (Ed.), *Un siécle d'electricité dans le monde* (pp. 7–22). Presses Universitaires de France.
Kaijser, A. (1994). *I fädrens spår. Den svenska infrastrukturens historiska utveckling och framtida utmaningar*. Carlssons.
Kaijser, A. (2002). Systembuilding from below. Institutional change in Dutch water control systems. *Technology and Culture, 43*(3), 521–548.
Kaijser, A. (2003). Redirecting infrasystems towards sustainability: What can we learn from history? In A. Biel, B. Hansson, & M. Mårtensson (Eds.), *Individual and structural determinants of environmental practice* (pp. 152–179). Ashgate.
Kates, R. W., Colten, C. E., Laska, S., & Leatherman, S. P. (2006). Reconstruction of New Orleans after hurricane Katrina: A research perspective. *Proceedings of the National Academy of Sciences, 103*(40), 14653–14660.
Keirl, S. (2006). Ethical technological literacy as democratic curriculum keystone. In J. R. Dakers (Ed.), *Defining technological literacy: Towards an epistemological framework* (pp. 81–102). Palgrave Macmillan.
Mays, L. W., Koutsoyiannis, D., & Angelakis, A. N. (2007). A brief history of urban water supply in antiquity. *Water Science and Technology: Water Supply, 7*(1), 1–12.
Melosi, M. V. (2008). *The Sanitary City: Environmental services in urban America from colonial times to the present*. University of Pittsburgh Press.
Mindell, D. A. (2002). *Between human and machine: Feedback, control, and computing before cybernetics*. The Johns Hopkins University Press.
Nia, M. G., Harandi, M. F., & de Vries, M. J. (2019). Technology development as a normative practice: A meaning-based approach to learning about values in engineering—Damming as a case study. *Science and Engineering Ethics, 25*, 55–82.
Nye, D. E. (2006). *Technology matters: Questions to live with*. The MIT Press.
Rees, W. E. (1997). Is "Sustainable City" an oxymoron? *Local Environment, 2*(3), 303–310.
(2007). *Standards for technological literacy: Content for the study of technology* (3rd Ed.). International Technology Education Association.
Sörlin, S. (2019). *Till bildningens försvar. Den svåra konsten att veta tillsammans*. Natur & Kultur.
Torbrand, D. (1978). Kommunikationernas förändring. In S. Hellström (Ed.), *Linköpings historia 4. Tiden 1863–1910*. Linköping.

Vaidhyanathan, S. (2018). *Anti-social media: How Facebook disconnects us and undermines democracy*. Oxford University Press.

Vermaas, P., Kroes, P., van de Poel, I., Franssen, M., & Houkes, W. (2011). *A philosophy of technology: From technical artefacts to sociotechnical systems*. Morgan & Claypool Publishers.

Wiener, N. (1948). *Cybernetics: Or control and communication in the animal and the machine*. Hermann & Co., Editeurs.

Wiener, N. (1950/1989). *The human use of human beings: Cybernetics and society*. Free Association Books.

Williams, P. J. (2017). Critique as a disposition. In P. J. Williams & K. Stables (Eds.), *Critique in design and technology education* (pp. 135–152). Springer.

Ynnerman, A., Löwgren, J., & Tibell, L. (2018). Exploranation: A new science communication paradigm. *IEEE Computer Graphics and Applications*, 13–20. Retrieved from www.computer.org/cga

Jonas Hallström is Ph.D. and Professor of Technology Education at the Department of Behavioural Sciences and Learning, Linköping University, Sweden, where he is also currently Deputy Head of Department. He is also Chair of the Swedish National Graduate School in Science and Technology Education Research (FontD). He presents regularly at international conferences and consults on technology education. He is one of the editorial board members of the Brill/Sense book series International Technology Education Studies. His research primarily concerns the historical emergence of technology as knowledge content in the school, the epistemology and subject philosophy of technology, various subject content (e.g. technological systems) as well as the attitudes to and knowledge of technology and technology education of students, student teachers and teachers. The research also relates to technology teaching in relation to e.g. design, gender (girls and technology), authentic learning, models and modelling, and STEM (science, technology, engineering, mathematics) education.

Arne Kaijser is Professor Emeritus of History of Technology at KTH Royal Institute of Technology, Stockholm. His main research interests concern the historical development of infrastructural systems from a transnational perspective. His latest books are: (with Per Högselius and Erik van der Vleuten) Europe's Infrastructure Transition. Economy, War, Nature (Palgrave, 2016); (editor with Markku Lehtonen, Jan-Henrik Meyer and Mar Rubio-Varas) Engaging the Atom: The History of Nuclear Energy and Society in Europe from the 1950s to the Present (West Virginia University Press, 2021).

Chapter 2
Technical Systems and Technical Artefacts: How to Conceptualize Their Relation

Marc J. de Vries

Abstract In technology education, technical artefacts and technical systems are terms that we frequently encounter. But what is the difference between the two? How are they related? In this chapter the two notions will be compared as two different perspectives that are complementary in our understanding of the human-made world. In the artefact perspective, the focus is on the (to be) physical object that designers can take decisions about and in the system perspective the physical object is seen as part of a larger whole of physical and non-physical and human-made and non-human-made elements. A theoretical framework for analysing aspects in reality developed by the Dutch philosopher Herman Dooyeweerd will be used to show how both perspectives can be brought together in an overall view on humans and their interaction with human-made objects.

2.1 Introduction

This book is about teaching and learning about technical systems. The notion of systems is an important one in technology education. It is a broad notion that overarches very different domains in technology. Also it connects technology to society. Hence, we often see the term socio-technical systems, to indicate that we do not treat technical systems as purely 'technical', but as also having a social dimension (Hughes, 1989). The notion of technical systems is closely related to another important notion in technology, namely that of technical artefacts. In this chapter we will explore the relation between those two notions but also offer an alternative approach in which the non-technical aspects are included in a different way than in the notion of socio-technical systems. In order to get a picture of that relation and its alternative we will call in the analytical philosophy of technology. This type of philosophy aims particularly at clarifying concepts. For educational purposes that is very relevant. How can we teach and learn concepts, of which the content and meaning is unclear

M. J. de Vries (✉)
Delft University of Technology, Lorentzweg 1, Delft, Netherlands
e-mail: M.J.deVries@tudelft.nl

J. Hallström and P. J. Williams (eds.), *Teaching and Learning about Technological Systems*, Contemporary Issues in Technology Education,
https://doi.org/10.1007/978-981-16-7719-9_2

and differs for each individual teacher and learner? Analytical philosophy on the one hand tries to do justice to the way we use terms in daily life, but sometimes also points out inconsistencies in that way of using terms and proposes a more consistent way. Both systems and artefacts are terms that we use in daily life and certainly in technology education, and we all have preconceptions about the meaning of those terms. Probably our intuitive use of the terms focuses on the material aspect of both. If we have a somewhat broader idea about these notions, perhaps we will ascribe social aspects to systems, but artefacts perhaps remain mainly material to most of us. In this chapter we will argue that the difference between technical artefacts and technical systems is not in having or not having a social dimension. But what, then, is a more proper way of conceptualizing the relation between technical artefacts and technical systems?

2.2 Non-Distinguishing Issues

The word 'artefact' literally means: 'made by craft'. Artefacts are human-made, and this distinguishes them from clouds, rocks, plants and animals, just to mention a few natural objects. Artefacts are made of material that ultimately has its origin in nature, but that material has undergone some kind of treatment to make it an artefact. In the philosophy of technology, this is further specified by identifying an artefact as an object that does not only have a physical 'nature' like all objects, but contrary to natural objects it also has a relational dimension (also called a functional nature) (Kroes & Meijers, 2006). To be an artefact not only requires being made by humans but also being ascribed a function. The notion of function then needs to be taken broadly. My relation to the artefact can also be that I ascribe aesthetic or economic values to it. When I limit this functional nature to useful functions, it is distinguished from wood chips that result from sawing a plank. They were man-made indeed but without a specific intention (although later a function can be ascribed to them as fuel material). Note that in this perspective, the physical object is analysed separately from its environment. It is acknowledged that the physical object does not stand on itself (i.e. in the very nature of a technical artefact) but for certain purposes (for instance, when designing the object) it can be helpful to focus on the physical object qua object and design it as such, thereby taking into account its functional relationships with the environment as constraints but not including the environment in the design.

Much of this can also be said about technological systems. Technological systems are also human-made and distinct from natural systems, such as ecosystems. Technological systems, too, have a physical and a functional nature. Here, there is no difference between artefacts and systems. What, then, is that difference? Let us explore two options.

2.3 Two possible ways of distinguishing a technical artefacts perspective from a technical systems perspective

An intuitive distinction between a technical system and a technical artefact could be that the notion of technical systems emphasizes the interaction between different elements in the object so much that one tends to include the interaction between the technical object and its human and social environment in the technical system almost 'by nature'. In that case, the technical system goes beyond the technical artefact as it was created by humans. This extension is possible because there is flexibility in defining the boundaries of the system. That holds even within the physical object. That is why engineers use the notion of sub-systems that are systems in themselves also. The engine is part of a car but it can be treated as a system in itself. The car as a system can also be seen as part of an even larger system that also comprises people and organizations. In that perspective systems are in fact always socio-technical systems. In the literature on technical artefacts, however, usually the technical object is seen as distinct from its environment. This could indeed be a legitimate way to make a formal distinction between technical systems and technical artefacts. Technical systems are socio-technical in nature, and technical artefacts are the physical part of technical systems. In that perspective there is the whole (socio-)technical system of road transportation and the car as a technical artefact within that.

Can one make the extension from the technical to the socio-technical also with the notion of technical artefacts? Yes, and that indeed happens in literature. Yet, this can cause some conceptual confusion. The literal meaning of artefact is: made by craft (usually taken to be also: made by humans, although one could also argue for spider webs and beaver dams to be artefacts as well). But how about the extension to socio-technical artefacts, including perhaps elements that are not human-made (such as the humans themselves or nature)? The overall 'technical artefact' would then have human-made elements and non-human-made elements. That seems to be not very logical. This observation supports the idea that the distinction between technical artefacts as only the physical object and technical systems as the totality of objects and their environment as given in the previous paragraph makes sense indeed. This does not take away, of course, that technical artefacts also make sense only in a broader social context.

At the same time, this distinction has some weakness. There are important socio-technical aspects that are not captured in this notion or only in a rather indirect way. Let us take moral considerations for example. One could claim that they come in through the humans in the socio-technical system. But in the analysis of the system, moral considerations cannot easily be deduced from the behaviour of people. Besides that, they are not distinguishable entities in the system. In a systems diagram they would not show up. Another weakness becomes clear when we see this notion from a designer's perspective. Someone designing a car will likely take into account all sorts of factors in the environment but still treat the car as a technical object only because that is what (s)he is designing, not the entire socio-technical system. Making a clear distinction between the technical object and its environment is much more fruitful than thinking from the immensely complex totality of the car and everything it relates with. This leads to another weakness of the notion of socio-technical systems: in the end it comprises the totality of reality, as ultimately everything is connected to everything. The car is related to its users. They are related to their work as their salary allows them to buy and use their car. Their work is related to the national economy, which in turn is related to international economies and etcetera. Of course, one can set a boundary to the socio-technical system at any level, but what legitimizes the choice of the boundary?

If we would confine the notion of technical system to the physical object rather than taking all technical systems to include non-technical elements, as with technical artefacts, there is another way to differentiate between a technical system and a technical artefact, namely on the basis that systems by definition consist of different parts that work together (the original Greek word sustèma was derived from 'sun', together', and istèmi', 'put together'), whereas this condition does not necessarily apply to technical artefacts. Technical artefacts can have one part only. One could, however, question the importance of that distinction. Systems inherently have a certain complexity because they consist of more than one part. The more parts they consist of, the more complex the system. There is, however, not a linear relation between the number of parts and the complexity, because some systems have many parts but many of them have the same function, while in other systems each component has a different function. All this can be said of artefacts too. Suppose we can make a car that consists of one part only. It would then in our current distinction be a technical artefact but not a technical system. But does that really matter for its relation to the environment? Maybe users would not even notice the fact that the car has one very complicated part only.

2.4 An Alternative Conceptualization

Both perspectives make sense and have weaknesses. The artefact perspective is particularly useful for the designer as that is what (s)he can manipulate. The systems perspective is a more natural way to take into account the relation between the artefact and its environment. Is there perhaps a way to combine both perspectives? The proposal I will make here is to view the car, in our recurrent example, as a physical object that can be regarded from different perspectives, both in case it has one part and in case it has multiple parts. Let us now turn to an analytical framework developed by the Dutch philosopher Herman Dooyeweerd. According to this framework, all objects (natural as well as human-made) exist in a reality in which many different aspects can be recognized. All objects exist in, for instance, what Dooyeweerd calls the numerical aspect of reality. Objects have something 'countable'. We can count the number of trees in a wood, but also the number of parts in a car. The numerical aspect of reality is related to the physical nature of artefacts and systems. It does not require a human for the artefact to have a certain number of parts, nor does it require a human for a collection of artefacts or systems to consist of a number of artefacts or systems. It does, of course, require a human to build a car with that number of parts, and later to count the number of parts, and thus turn this aspect into something more than just theory by using the knowledge of the number for calculating costs, for instance. But the property of having a number of components in itself does not depend on a human being there to count it. It is intrinsic, not relational. We see that this aspect provides insights relevant for both possible distinctions between technical artefacts and technical systems. For the distinction based on number of parts (any for an artefact but only more than one for a system) the numerical aspect is crucial. But we also noted it helps us understand that the social element is not in the system in the same way as the number of parts because of the difference between relational and non-relational aspects. So we will treat the notion of technical artefact and the notion of (socio-)technical system as two perspectives that are complementary and brought together in the multi-aspectual approach.

Let us now turn to another of the aspects of reality as identified by Dooyeweerd, the social aspect. The idea here is that, apart from being a numerical entity (see the previous paragraph) a technical artefact is also a social object. That is, it functions as something that is part of the interaction between (groups) of people. Clearly this has a lot of practical meaning for technical artefacts/systems. This is the very reason why the term socio-technical system has emerged. This perspective may not be the most practical for the design of a new object but it has great merit when considering the social impact of a new technology. But also this perspective can help designers and engineers to design in such a way that the new technology is likely to fit well in society in all its complexity. The socio-technical systems perspective has had quite some implications, for, e.g., the bike. Technically this device allows people in society to move from A to B. But why one would like to use it can be very different. Originally the bike was seen as a macho machine for boys to show to the girls how brave they were. A very large front wheel and a very small wheel were quite suitable for that. But when the social view on the bike changed to a transportation means that women and girls could also use, two equal wheels were much more convenient (Bijker, 1995).

The numerical and the social aspects are two of the fifteen aspects of reality that Dooyeweerd has identified. Table 2.1 presents the whole series. For each of the aspects an example from technology has been added a well as an example of a ‘law’ or ‘rule’ that holds for that particular aspect. All entities in reality will somehow obey these ‘laws’ or ‘rules’. Note that for the first five aspects these laws are obeyed by natural necessity and for the other aspects it is up to humans to obey them or not. In other words, the laws in the first five aspects are descriptive and for the remaining aspects they are normative/prescriptive.

Having reflected on the numerical and social aspect, we will now see how the other aspects help us understand the nature of technical artefacts and technical systems and the relation between these two notions. After all, this understanding is important when we want to teach about artefacts and systems.

Continuing from the numerical aspect we get to the spatial aspect. From the technical artefact perspective the meaning of this aspect is clear: each physical object takes space and so does a technical artefact. This is something the designer has to

Table 2.1 Aspects of reality in Dooyeweerd's ontology (Dooyeweerd, 1955; Verkerk et al., 2016)

Name of aspect	Relevance for technology (example)	Regularity (example)
Numerical (quantitative)	Number of parts	$1 + 1 = 2$
Spatial	Space taken by product	Sum of angles in triangle = 180 degrees
Kinematic (motion)	Joint can move (or not)	Connection with glue do not allow relative motion of parts
Physical	Material properties	Conservation of energy
Biotic (life)	Contact with living tissue	Survival of the fittest
Psychic (sensitive)	Customer perception of product	Psychological consumption law
Analytical (logical)	Reasoning about the product	Rules of logical inference
Formative (historical)	Product development process	Logic of steps in design
Symbolic (linguistic)	Name of the product	Grammar rules
Social	Impact of product on society	Politeness
Economic	Cost of product, price of product	Law of diminishing returns
Aesthetic	Appearance of product	'Simplicity is beautiful'
Juridical	Compliance with legislation	Private or public laws
Ethical	Safety or sustainability of product	Moral code in company
Belief (trust)	Trust evoked by the product	'Trust comes on foot and goes on horseback'

take into account because the artefact will have to fit in an available space. The presence of the technical artefact will have consequences for other objects in its direct environment because every space can only be occupied once. Folding or squeezing the artefact may save space, but there is always a minimum needed space. The number of parts does not seem to play a vital role here. A one-part artefacts can take more space than a multi-part artefact as well as less space. How about the (socio-)technical system perspective? This perspective has an added value because it makes us realize the connections to other aspects. Generally speaking, space mean money (in the case of storage boxes and warehouses for instance). Objects taking space may also have an impact on how people perceive the environment. Wind turbines do not just take space but by doing so they change the landscape, which may make people living there not happy. So both perspectives help us understand what conditions the spatial aspect poses to the design and use of the device.

Next we have the aspect of motion. Things move in space, even when the motion is almost unnoticeable. Motion can be needed for functioning or is to be minimalized. Skyscrapers do not move forward but they do swing in the wind, and designers have to take that into account. Famous is the Tacoma Bridge that was not meant to move but was destroyed by the continuous swinging caused by the wind. There is a relation within the spatial aspect because the motion of an artefact means that it takes more space than just its own size, again something to be taken into account when designing the object. Again the aspect has a practical meaning in the technical artefact perspective. The number of parts is relevant in that for multiple-part artefacts, the designer has to decide if and how parts move relative to another. The (socio-) technical systems approach brings in the connections to other aspects. Shaky objects may influence the trust people have in a negative way. Who dares to cross a bridge that is swinging? Again both perspectives are complementary for our understanding of the relevance of the aspect.

Then comes the physical aspect. This comprises the domains studied primarily by physics and chemistry. Keywords are energy and physical interaction. The technical artefact perspective points out the importance of physical/chemical interactions within the artefact and between the artefacts and its environment. Bridges exert forces on the ground and wind and rain exert physical and chemical influences on the bridge. The number of parts is relevant here in that multi-part artefacts will probably have more internal physical interactions (friction, for instance) than a one-part artefact. All that has to be taken into account when designing the bridge as a technical artefact. The (socio-)technical systems perspective brings in the consequences of these interactions for humans and society. The use of a bridge is limited by the maximum forces it can take and this limits the amount of traffic passing over it, which has economic and social consequences. A bridge can have a strong social effect when it connects communities (like the famous chain bridge that connected the villages Buda and Pest and turned them into the city of Budapest).

The biotic aspect at first sight may not be so relevant for technical artefacts (unless they are made of living material of course). This is, however, not the case. Technical artefacts often interact with humans, animals and/or plants. Small organisms can grow on the bridge and cause deterioration. Humans walk over the bridge, touching the handrail. Cars produce gases that are inhaled by people. All interactions between a technical artefact and organisms it interacts with need to be taken into account by the designer. Here the number of parts does not seem to be distinctive. The socio-technical systems perspective points to the fact that the interaction between technical artefacts and their environment gives rise to all sorts of questions related to other aspects: artefacts potentially causing illness will have an economic impact, a trust impact, a moral impact, etc.

In the psychical aspect the technical artefacts can only be an object, as it cannot perceive anything by itself (whereby I reckon the phenomenon of purely physical forms of 'perception' like light falling on a charge-coupled device (ccd) part of the physical domain and not of the psychic domain). Note that we have now passed the 'border' between the aspects directly related to the physical nature of the artefact (the non-intentional aspects in which the artefact can function as a subject as no intentionality is needed for that) and the 'higher', intentional aspects (in which the artefact can only be an object as it has no intentionality itself). The name of this aspect immediately reminds of the title of a well-known book on design: The Psychology of Everyday Things by Donald Norman (Norman, 1988). This book nicely spells out how the expected perception of the artefact by its user is quite important for the designer. The appearance of the artefact can either confuse or inform the user. In this aspect the number of parts can be relevant as one-part artefacts can look simpler than multi-part artefacts and particularly when the user has to distinguish between different parts with similar appearance that can be problematic. The socio-technical systems perspective makes us aware that perception is not only individually determined but also socially. When we see a coin on the street, we perceive more than a piece of metal because of the socio-economic meaning of the coin.

The analytical aspect of reality is about the opportunities reality offers to distinguish. We can tell black from white, good from evil, true from false, etc. This also applies to technical artefacts. We can tell from well-functioning from malfunctioning, we can tell simple from complex. We can tell one-part from many-part artefacts. In fact our whole discussion about the two perspectives that are dealt with in this chapter is a matter of analysis. Every science has a foundation in this aspect, including philosophy. The (socio-)technical systems perspective makes us aware of the fact that also what we include in our analysis and how we do that is at least partly socially determined. In the modern Western world we use different types of logic than in other parts of the world. That also applies to the way we analyse technical systems.

The formative aspect can also be called the technical aspect as it deals with the fact that reality offers opportunities for creative acting by humans. It could also be called the cultural aspect. We shape things: artefacts, poems, laws, moral codes and religious expressions. For this aspect there are 'laws', that is there is a certain logic in developments. Design is always re-design in some way or other and there is a logical step in the line of development. French philosopher of technology Gilbert Simondon found this so important that for him a technical artefact is not a thing but a process of development (De Vries, 2008). A car in his view is not the thing on the road but a line of development starting from steam engine driven cars to the electrical car of today. Some of that logic is the same when seen from a technical artefact perspective or from a (socio-)technical systems perspective, but some of that logic is particular for either of those perspectives. A multi-part artefact calls for different decisions than a one-part artefact (related to the way parts have to be connected and interact). This aspect therefore has a special meaning for our understanding of the difference between the perspectives.

The symbolic aspect refers to symbols related to the technical artefact or system. This can be in the name (e.g., 'screwdriver'), in the manual, but also in the shape. Some door handles by their shape clearly communicate: push me, while others are clearly meant for pulling. Red buttons mean: don't push me lightly and green lights mean: push me to get what you want. Clearly this has a meaning for the design of the technical artefact qua physical object, but the meaning of the symbol cannot be understood but from a socio-technical systems perspective. The meaning of red and green is socially determined. Again we see how the aspect makes us consider both perspectives as complementary insights.

We skip the social aspect because it was discussed earlier already, and go to the economic aspect. The designing, making and using of technical artefacts costs money and produces money. The technical artefact aspect focuses on the costs related directly to the physical object and the socio-technical systems perspective on the way society determines the economic value of the physical object. Again both perspectives contribute to our understanding the meaning of that aspect in their own way.

We are now in the realm of value ascription to objects and their social function. We had social value in the social aspect, 'money' value in the economic aspect, and we now turn to values of beauty (or ugliness) in the aesthetic aspect. What we regard as beautiful or ugly is at least to some extent socially determined (in some cultures slim women and considered beautiful and in other cultures fat women are). Designers have to translate vague and generic notions of beauty into concrete shapes and colours. Again both the artefact and the system perspective are needed to see the meaning of this aspect of reality.

Next are values of just and unjust that feature in the legal aspect. What has been said about the aesthetical aspect easily translates to this aspect. The same holds for the aspect that is concerned with values of (morally) good and bad in the ethical aspect and in the values of trustworthiness or suspiciousness.

It is good to emphasize that this division into fifteen aspects is of course a choice and alternative choices are possible. The nice thing about fifteen aspects is that they are in between only two (as in the dual nature of technical artefacts approach; see Kroes & Meijers, 2006) and a hundred (that no doubt we would be possible to come up with). Besides that, fifteen was sufficient to illustrate the idea. Using the multi-aspectual approach enables us to combine a technical artefact and a (socio-) technical perspective. To summarize the above, Table 2.2 shows how the artefact and the systems perspective feature in each of the 15 aspects. For each aspect one example is mentioned. In any concrete case the second and third column can be given a certain content.

2.5 Implications for Technology Education

What relevance does all this have for education? There are other chapters in which this is spelled out in detail, but here some hints will be given. In the first place it is useful that pupils learn to reflect on the nature of artefacts and systems because this can support the way they design, make and use them (which happens all the time in technology education). The social dimension of the simplest artefact (like a coin) and of the most complex system (like an airplane) should be taken into account by learners for a good understanding of what technology is all about. Furthermore, learners should gain some insight of the consequences of making an artefact in which one part fulfils the overall function that it was designed for, or to design and make different parts for different sub-functions. Designing is a very appropriate way of learning the notions of artefacts and systems and therefore the approach of design-based concept learning can be recommended here. Designing artefacts and systems forces the learner to make conscious decisions about the device and thus stimulates careful reflection on their nature.

Both the technical artefact and the (socio-)technical systems perspective can be part of technology education practice. The extent to which the full complexity as

Table 2.2 Artefact and system perspectives combined in the aspect approach (examples)

Name of aspect	Artefact perspective	Systems perspective
Numerical (quantitative)	Number of elements in physical object (1 or more)	Number of elements in the whole system (always more than 1 physical, plus non-physical elements)
Spatial	Space the object takes	Social appreciation (horizon 'polluted')
Kinematic (motion)	Physical motion of physical object (shaking)	Social appreciation (not safe)
Physical	Material properties	Social perception (plastic? not good for the environment!)
Biotic (life)	Living beings are not always part of the artefact (external relation only)	Living beings are part of the socio-technical system
Psychic (sensitive)	Perception of the physical object (a small round metal piece)	Socially determined perception (a coin)
Analytical (logical)	Internal logic of the design	Socially determined 'logic' (technically it may function, but socially it may not)
Formative (historical)	Design work leading to the physical artefact	Re-design by society (e.g. from conception to anti-conception pill)
Symbolic (linguistic)	Physical realization of symbol ('green')	Socially determined meaning of symbol ('safe')
Social	The artefact is shaped according to social needs (a bike has equal wheels or very unequal wheels, depending on social constraints)	Determining aspect for how the product is seen (macho machine or transportation means)
Economic	Price of materials of artefact	Price society is willing to pay for the product
Aesthetic	External characteristics of the artefact (e.g. curved or strait)	Socially determined aesthetic appreciation
Juridical	External characteristics of the artefact (hat designed to indicate a police officer)	Socially determined (in)justice appreciation (that woman is always after you)
Ethical	External characteristics of the artefact (coat designed to indicate a nurse)	Socially determined moral appreciation (that is a good lady caring for people)
Belief (trust)	External characteristics of the artefact (robe indicating a pastor)	Socially determined trustworthiness appreciation (to that man you can trust your religious concerns)

spelled out in the fifteen aspects of reality (according to Dooyeweerd) is used can depend on the educational level. In primary education it can be reduced to the physical aspects (related to the physical or non-intentional nature of the artefact) and the functional aspects (all the aspects that involve intentionality). This division in two is strong as it helps pupils distinguish between the physical make-up of the artefact ('structure' is a term that can be used for that) and the function (taken broadly to include also what consumers are willing to pay, aesthetic appreciation, etcetera). In a design activity they can thus learn to see the difference between a need and a solution. In a list of requirements, something like 'should be made of wood' is not really a need. It is a solution that has already been taken for granted without realizing that the underlying need can perhaps be addressed by an alternative choice that does more justice to other requirements. So even the simple two-fold distinction already contributes significantly to the pupils' understanding of technology and design. Here the technical artefact perspective can show its value without necessarily being complemented by a (socio-)technical systems perspective.

In higher levels of education both sides can be spelled out in more detail. To the physical nature belong size and shape, number of parts, physical and chemical material properties and relations with living organisms. The functional nature can be spelled out as, for instance, economic and social issues, but also aesthetic, legal, moral and trust issues. Thus pupils gradually discover that design is truly challenging as the artefact will have to function in all aspects and therefore is subject to all 'laws' that hold in those aspects and this leads to a comprehensive list of requirements with various possible conflicts between requirements (price versus beauty or versus complying with environmental legislation). Representing each aspect with its own icon can be helpful for pupils (for an example, see https://thinkfaith.net/fisch/blog/multi-faceted-meaning-life). The more the aspects come into play, the more important the socio-technical systems perspective will help understand that complexity.

Once more we have here an example of how philosophy of technology can support technology education (De Vries, 2018). Philosophy tends to bring out the essence of things and get away from all the details that make things complicated or map out the complexity to make it more comprehensible (as in the Dooyeweerd approach). And is that not what helps us in education? We do not want to confront learners with the full complexity of reality immediately. Philosophy helps us to tell what to start with and what to bring in at a later stage. Hopefully this chapter contributes to that.

References

Bijker, W. (1995). *Of Bicycles, bakelites and bulbs. Towards a theory of sociotechnical change.* MIT Press.

de Vries, M. J. (2008). Gilbert Simondon and the dual nature of technical artifacts. *Techné, 12*(1), 23–35.

de Vries, M. J. (2018). Philosophy of technology: Themes and topics. In M. J. de Vries (Ed.), *Handbook of Technology Education* (pp. 7–16). Springer.

Dooyeweerd, H. (1955). *A New Critique of Theoretical Thought, Volumes I-IV*. Paideia Press.

Hughes, T. P. (1989). The evolution of large technological systems. In W. E. Bijker, T. P. Hughes & Pinch. T. (Eds.), *The social construction of technological systems. New directions in the sociology and history of technology* (pp. 51–82). MIT Press.
Kroes, P. A., & Meijers, A. W. M. (2006). The dual nature of technical artefacts. *Studies in History and Philosophy of Science, 37*(1), pp. 1–4.
Norman, D. A. (1988). *The psychology of everyday things*. Basic Books.
Verkerk, M. J., Hoogland, J., van der Stoep, J., & de Vries, M. J. (2016). *Philosophy of technology. An introduction for technology and business students*. Routledge.

Marc J. de Vries is Professor of Science and Technology Education and Professor of Christian philosophy of technology at Delft University of Technology, the Netherlands. He is the editor-in-chief of the *International Journal of Technology & Design Education* and the coordinator of the international conference series Pupils' Attitudes Towards Technology (PATT). He was the author of the book *Teaching About Technology: An Introduction to the Philosophy of Technology for Non-Philosophers* and the book *80 Years of Research at Philips*. He was the editor of the *International Handbook of Technology Education*, published by Springer.

Part II
Contents, Concepts, and Contexts for Teaching About Technological Systems

Chapter 3
Technological Systems in National Standards and Curricula

Per Norström

Technological systems consist of interacting components that contribute to a common function. This chapter presents a description of how systems are included in national curricula and standards for technology education intended for use in primary and lower secondary school in Australia, Europe, New Zealand, South Africa, and the United States. Systems are seldom identified as a separate strand or theme – they are learnt in conjunction with other technological phenomena. In curricula that emphasize designing and making, included systems are those that pupils can design and make from simple components and/or prefabricated subsystems. In curricula that include the social aspect of technology, infrastructure and other large systems tend to have more prominent positions. Cross-disciplinary 'systems thinking' is mainly included in curricula where technology is combined with the natural sciences. It should be noted that the roles of non-technical components (such as institutions or human agents) in technical systems are not highlighted in the studied curricula.

The purpose of this chapter is to describe and discuss how and to what extent systems and systems thinking are incorporated in national curricula and standards in technology education. The term 'system' is used in a wide variety of ways. The first question that must be resolved is therefore: What definition of *system* should we use to capture relevant aspects of teaching and learning when discussing technology education?

In his seminal work, *General System Theory* (1973, p. 33), Bertalanffy states that systems are 'complexes of elements standing in interaction'. Further, IEEE (2007, p. 7) describes a system as: 'A set or arrangement of elements [people, products (hardware and software) and processes (facilities, equipment, material, and procedures)] that are related, and whose behaviour satisfies operational needs and provides for the life cycle sustainment of the products'.

Fundamentally, and uncontroversially, systems are related to a set of interacting parts (elements or components). However, describing the relation between actual objects and systems does not attract universal agreement. In the two previously

P. Norström (✉)
KTH Royal Institute of Technology, Stockholm, Sweden
e-mail: perno@kth.se

J. Hallström and P. J. Williams (eds.), *Teaching and Learning about Technological Systems*, Contemporary Issues in Technology Education,
https://doi.org/10.1007/978-981-16-7719-9_3

mentioned examples, the system is described as a set, complex or aggregate of elements that taken together make up the system. This is not the only possible viewpoint. Magee and de Weck (2002, p. 4) state that a system is 'a set of interacting components having well-defined (although possibly poorly understood) behaviour or purpose; the concept is subjective in that what is a system to one person may not appear to be a system to another'. It should be noted that they state that it 'is a system', not 'appears as a system'. According to this description, whether something is a system is at least partly dependent on the observer; how the components appear to them affects their systemic identity. Olsson and Sjöstedt (2004, p. 9) take this concept even further, stating that 'systems are mental constructs or models of a specified part of reality'. For them, the model or mental construct *is* the system, meaning that the system only manifests when an onlooker forms a mental construct or creates a model of the interacting components and regards them as a system.

Hereafter, the term *system* will be used to describe the actual complexes of objects, which is broadly in accordance with Bertalanffy (1973) and IEEE (2007). The term *system model* will be employed for simplified representations of systems (which are referred to as systems by Olsson & Sjöstedt, 2004). This appears to be in agreement with how systems are understood in most curricula and standards.

3.1 Small and Large, Technical and Socio-Technical

Systems can be large or small, complicated or simple. Moreover, almost all non-trivial technical artefacts consist of more than one part, meaning that they could be considered systems (see also Chap. 2 "Technical Systems and Technical Artefacts: How to Conceptualize Their Relation" by Marc de Vries in this book).

For example, the stapler on my desk definitively exhibits systemic characteristics, even though it is rather small and has limited functionality. By comparison, the international air transport system is vast and consists of innumerable components. Some of these components are physical objects (such as aeroplanes, runways, and fuel depots), while some are social artefacts (such as laws, rules, standards, and money). Importantly, some components are human, such as pilots, pursers, luggage handlers, and mechanics. System sizes and the roles human beings play in systems are of course continua. Some systems work without any human interactions (at least after being switched on), while others (such as the stapler) require a human user to provide the requisite force. If we regard the car as a system, the driver acts as sensor and actuator, becoming an important part of the control system. In air traffic systems, both humans and institutions play important roles. Moreover, although aeroplanes could in principle fly without safety standards or insurance, the air traffic system we know today would be impossible. In small, deterministic systems such as the stapler, all components lack will, intentions, and an awareness of objectives, which is not the case in large systems (e.g. Kroes et al., 2006).

Even though the stapler and the international air traffic system are considered systems, it is not easy to find similarities concerning how they could be studied in school or what could be learned from them. The stapler could be used to provide examples of engineering mechanics, ergonomics, industrial design, materials science, standardization, and the interaction of mechanical components. By comparison, the international air traffic system could provide teaching examples about logistics and large-scale planning, communication, risk management, and environmental effects. Accordingly, although both are systems, they tend to be studied in different ways and provide very different opportunities for learning.

When analysing small, deterministic systems such as the stapler, theories from classical physics combined with a knowledge of ergonomics and some engineering know-how suffice. For more complex systems that involve feedback loops, classical systems theory can be employed, which is a highly mathematised discipline based on areas including control theory and cybernetics. However, for larger systems where people, money, laws and a huge number of ill-defined inputs and outputs interact in complex non-deterministic ways, this is not enough. While the analysis of these systems, which are commonly referred to as 'socio-technical systems' or simply 'large technological systems', relies on theories and methods from engineering, other disciplines are required too, such as sociology, decision theory, law, and behavioural sciences. The social and technical aspects are closely intertwined. Moreover, when only considering the physical technical artefacts that are part of a system, these are referred to as its 'technical core' (e.g. Hughes, 1983).

As will be shown later, both large and small systems are included in curricula and learning standards. However, small systems tend to dominate in curricula based on designing and making, while larger systems are mainly considered in curricula that include social, political, and historical issues.

3.1.1 Systems Thinking

'Systems thinking' relates to the ability to discuss, analyse and understand a system as a system, which includes creating models (mainly mental, physical, or symbolic) and drawing conclusions about individual parts and the entirety. In addition to technology education, systems thinking is discussed in engineering (e.g. Frank, 2002), biology (e.g. Rosenkränzer et al., 2017), and geography (e.g. Schuler et al., 2018). Although details differ, the main building blocks of systems thinking appear to be cross-disciplinary: to be able to study individual parts and the whole, to identify connections and causality, and to create models (mental or other) useful for analysis, prediction, problem-solving, and for revealing the systems' inner workings.

When describing the *technologies* learning area in the Australian curriculum (ACARA, n.d., p. 6), systems thinking is portrayed as follows: 'Systems thinking is a holistic approach to the identification and solving of problems where the focal points are treated as components of a system, and their interactions and interrelationships are analysed individually to see how they influence the functioning of the entire system'. Further, ITEA's *Standards for Technological Literacy* (2007) include a similar take on the concept: 'Systems thinking involves understanding how a whole is expressed in terms of its parts, and conversely, how the parts relate to each other and to the whole'. (p. 32), and 'Systems thinking is a practice that focuses on the analysis and design of the whole system as distinct from its many parts' (p. 38).

Related systems thinking concepts are presented in the National Research Council's *Practices, crosscutting concepts, and core ideas* of science education (2012). Here, a system is described as 'an organized group of related objects or components that form a whole' (p. 92). Multiple types of systems are mentioned (including ecosystems and the carbon dioxide cycle) in addition to common concepts such as inputs, outputs, and the necessity of bordering. Further, the use of models for planning, understanding, and predicting is emphasized. Overall, the National Research Council's views of what should be learnt about systems in science and technology education closely resemble the abovementioned 'systems thinking' concepts, even though the actual expression is not used.

The Massachusetts syllabus (Massachusetts Department of Elementary & Secondary Education, 2016) was explicitly inspired by the National Research Council (2012). The general nature of feedback systems is discussed in the document's introduction, where they are portrayed as playing similar roles in ecosystems, living organisms, and technical heating and cooling systems. The document states that '[a] focus on core ideas helps students to understand mechanisms and causes underlying a range of phenomena and apply their content understandings to real-world and novel situations.' (p. 9). Hence, although the syllabus expresses a belief in the existence of a kind of general and transferable systems thinking, once again, the actual term is not used.

It is notable that none of these descriptions mention that the possibility of different varieties of systems thinking. In many obvious ways, the mechanics-based systems thinking necessary to analyse the stapler is different from the systems thinking required to analyse the air traffic system. Further, I have not been able to find any studies indicating that proficiency in systems thinking concerning small systems is transferable to large ones, or vice versa.

3.1.2 What to Look for in the Curricula, Syllabi, and National Standards?

Based on the descriptions of systems and systems thinking in the previous sections, the following guidelines were used to identify content related to technological systems when scrutinizing the curricula, syllabi, and standards:

A phenomenon mentioned in the studied texts is regarded as a technological system if it

(1) *consists of two or more elements that interact and contribute to a common function,*
(2) *is predominately physical in nature,*

(3) *is designed by humans, with a certain purpose, function, or sub-function in mind, and*

(4) *is described either explicitly as a system (using the term 'system') or in another way that emphasizes its systemic nature (i.e. both as a functional whole and as an aggregate of components).*

Here, guideline (1) emanates from established definitions and descriptions of what characterizes systems. Guideline (2) is included to avoid discussions about the nature of technological artefacts. For example, whether a computer program is an artefact or a system in and of itself, or only when it is stored in a physical memory. Guideline (3) is typical of technology and is generally regarded as a defining characteristic of technical artefacts. It should be noted that some involved in the design might not be aware of the system; they may only be aware of the part on which they are working. Guideline (4) is demanded by the various concepts of systems thinking.

3.2 Selection of Documents

The included curricula, syllabi, and standards represent a convenience sample. The chosen documents are all easily available online and are written in any of the limited set of languages that the chapter's author reads (Danish, English, and Swedish). They are from Australia (ACARA, n.d., 2015), Denmark (Børne- og undervisningsministeriet, 2019), England (Department for Education, 2013), Finland (Utbildningsstyrelsen, 2014), Ireland (National Council for Curriculum and Assessment, 2016), New Zealand (Ministry of Education, 2018), Scotland (Scottish Qualifications Authority, 2012), South Africa (Department of Basic Education, 2011), Sweden (Skolverket, 2018), the USA (ITEA, 2007; Massachusetts Department of Elementary & Secondary Education, 2016; National Academy of Engineering, 2010; National Research Council, 2010, 2012, 2013), and Wales (Departement for Children, Education, Lifelong Learning and Skills, 2008, 2009). Although excluding large parts of the world in this way is clearly a coarse limitation, the included documents provide a broad sample. They are from different continents and represent different traditions in technology education.

The content and purposes of technology subjects vary considerably between countries. For example, only some include computer programming, and only some include the history of technology (Norström, 2016). Further, although most countries incorporate some kind of design or product development process, there are differences in the way this manifests. In some countries (e.g. Finland and Scotland), technology subjects are largely craft-based, while others (e.g. New Zealand and Sweden) have a broader focus.

The design of curriculum documents also varies. Some documents only contain a framework that allows the teachers to fill in the details, while others are detailed and provide teaching examples. Key Stage 3 in the English syllabus for design and technology (Department for Education, 2013) is an example of the former, while the corresponding Australian document (ACARA, 2015) accords with the latter. The English syllabus consists of three pages with a short introduction that is followed by a list of themes presented as bullet points. By contrast, the Australian document contains approximately 50 pages (including examples and a glossary), rendering comparisons between the syllabi difficult and not very useful. Accordingly, such comparisons are not a major feature of this chapter. The purpose is to provide an overview of different ways of including technological systems in curricula and standards.

Although technology curricula vary across the world, they tend to rely on three main traditions of knowledge for their content and justification of knowledge. Nordlöf et al., (2021) characterize these as *technical skills* (founded on the crafts and traditions of making, traditionally non-academic and largely based on experience), *technological scientific knowledge* (founded on engineering science and applied natural science), and *socio-ethical technical understanding* (founded on the subjects of history, sociology, ethics, and political science related to technology, or the academic disciplines *science, technology and society* or *science and technology studies* [*STS*]). Some technology curricula attempt to include all three knowledge traditions, while others are strongly dominated by one.

3.3 Method

Curricula, syllabi, and standard documents were studied, using a procedure inspired by the thematic analysis method of Braun and Clarke (2006) and the content analysis of Krippendorff (2013). Parts of the texts where technological systems (according to the previous description) are mentioned were highlighted, and the general directions for each curriculum were considered. The system-related text fragments were analysed using an inductive approach and grouped into themes based on what was present in the text concerning factors such as what kinds of systems were mentioned, what pupils should learn, and the terminology used.

3.4 Results

Technological systems are seldom identified as a theme or strand of their own in the curricula and standards. Instead, systems and systems thinking are parts of other themes, such as design or domestic technologies. This results in their being shaped by the general purposes, aims, and other content of related documents. This section begins with a general description of how systems-related content is included in curricula that are dominated by different knowledge traditions, followed by an overview of what pupils are expected to learn about systems.

3.4.1 Technological Systems Related to Traditions of Knowledge

Dominant knowledge traditions can significantly affect how and to what extent technological systems are included in standards and curricula. For example, craft-based curricula do not emphasize the same system characteristics as more science-based curricula. The study of systems is included in (and should supposedly fit into) a larger context.

3.4.2 Technological Systems in Curricula and Standards Dominated by **Technical Skills**

The Scottish subject of design and technology (Scottish Qualifications Authority, 2012b, p. 4) is a 'practical, exploratory and experiential' subject that emphasizes craft, design, and graphical experiences. Using Nordlöf's et al. (2021) terminology, it is strongly dominated by *technical skills*. The terms 'system' and 'systems thinking' are not used in the curriculum text. Moreover, the representation of technological systems (or lack thereof) in the Scottish curriculum is typical of technology subjects that rely on craft, sloyd or *technical skills* and have a strong design-and-make focus. The included systems are small and are primarily studied through making and direct manipulation, so although a form of systems thinking concerning small systems is implied, it is not elaborated upon.

Systems-wise, the Finnish subject of sloyd (Utbildningsstyrelsen, 2014), is similar to Scotland's design and technology. However, the syllabus does encourage visits to local industries, which could result in the inclusion of system-related content. The curriculum also encourages the use of electronic sub-systems in artefacts created by pupils.

Similar to its Scottish namesake, the Welsh subject of design and technology (Department for Children, Education, Lifelong Learning and Skills, 2008) is based around designing and making. However, the syllabus does include several examples of small technological systems (such as mechanical systems using levers and electrical circuits) and even mentions programmable systems such as floor turtles, but no large or socio-technical systems are included.

3.4.3 Technological Systems in Curricula and Standards with Explicit **Technological Scientific Knowledge** *Content*

The American *Standards for K–12 Engineering Education*, published by the National Research Council of the National Academies (2010), display a comparatively strong leaning towards engineering science and applied natural science. Design and product development work is referred to as 'engineering' and is described as an opportunity to provide a larger picture of technology, not just discrete facts. This could imply a level of systems thinking. In the standards, so-called *crosscutting concepts* are included. These are phenomena that require scientific, technological, and social perspectives to be fully understood. Addressing technological systems is one of the included concepts. For example, the dependence of modern civilisation on large technological systems is suggested as part of the content (National Research Council, 2010, e.g. p. 137; 2012).

Various state-specific American curricula rely heavily on the aforementioned standards. Examples include Vermont, New York State, and Massachusetts. In the Massachusetts curriculum (Massachusetts Department of Elementary & Secondary Education, 2016), the purpose of technology and engineering education is that pupils should learn about the opportunities and disadvantages of modern technology. The aim is to facilitate students' participation in the public debate and to encourage them to become careful consumers of technological products and information. Furthermore, education should prepare pupils for future learning and professional careers (pp. 10–11). Throughout the document, many examples of systems used in everyday life are provided. These include small ones (home cooling systems and house electricity) and large ones (transport systems and manufacturing systems). The use of mathematics and natural sciences (mainly physics) to analyse and understand systems is encouraged. Systems thinking is not mentioned specifically, but it is nevertheless included among the intended outcomes: abstract description of systems (p. 48), system models (e.g. pp. 72, 96, 110), and systems regarded as a whole and as parts (p. 56).

The Danish syllabus for science and technology (*Natur/teknologi* [da]) provides other examples of technological systems, which are mainly studied using natural sciences. An overarching theme of the subject is *technology and resources*, which includes the study of technological systems such as electricity, water, and sewage (Børne- og undervisningsministeriet, 2019, pp. 12, 41, 44).

The fact that modern society relies on large systems is noted repeatedly in both the American and Danish documents. However, this is not elevated to the same level as engineering science-based descriptions. Engineering science and natural science should contribute to pupils' understanding of technological systems, while the inclusion of some of the social sciences appears to be considered less important by the curriculum authors.

3.4.4 Technological Systems in Curricula Encouraging **Socio-Ethical Technical Understanding**

Among the studied curricula, none are dominated by the social and political aspects of technology and technological development. However, there are a few technology subjects that explicitly include *socio-ethical technical understanding* as one of several important perspectives. This is most obvious in Australia and Sweden.

In the Swedish syllabus for technology in primary and lower secondary school (Skolverket, 2018), technological systems are included in the stipulated core content throughout the nine years of schooling. The socio-ethical perspectives of systems are described explicitly for Years 7–9, when 'the benefits, risks, and limitations' of global technological systems are included in a larger theme of 'technology, man, society, and the environment' (p. 299). Among its aims, the Australian learning outcomes include 'make informed and ethical decisions about the role, impact and use of technologies in the economy, environment and society for a sustainable future' (p. 5). Further, it is also claimed that '[u]nderstanding the complexity of systems and the interdependence of components is necessary to create timely solutions to technical, economic and social problems' (p. 6).

3.5 What Are Pupils to Learn About Systems?

Different types of systems are studied according to variation between different curricula, standards, and intended learning outcomes. Moreover, whether pupils should be able to describe systems, create models of them, or something else also varies. The system-related content in the studied documents can be divided into the categories presented in the following sections. Note that the categories are overlapping and not mutually exclusive. One curriculum (or theme, strand, or element of core content within a curriculum) can include more than one of the themes.

3.5.1 Learning to Use Systems

Students should use objects that could be considered systems or parts of systems and be aware of their systemic characteristics. Typically, this takes the form of knowing how to use a computer system and having a working knowledge of what roles its different parts play. The pupil is predominantly identified as a user of systems, requiring only rudimentary knowledge of their systemic nature. This could be the first step towards learning to describe, model, or analyse systems (see below).

The New Zealand syllabus includes examples of this kind. For example, pupils learn to use computers and various software packages for creating digital web content. They also learn about the computer's parts, how information is stored, and gain 'understandings of how to build, install, and maintain computers, networks, and systems' (Ministry of Education, 2018, p. 4). Note that the focus is on *understanding* how to install and maintain the systems, not necessarily to actually do it. Pupils become qualified users who know their equipment, not network technicians. This is in line with the overarching purpose of technology studies in New Zealand, which is to become an informed citizen with a platform for future careers in technology.

Similar content is included in the Irish technology subject, where pupils' skills in using computers are emphasized. These skills incorporate basic usage, information retrieval, and using computers as design tools. They also learn to 'identify' hardware and 'understand system specifications' (National Council for Curriculum and Assessment, 2006, p. 19). The use of computer systems is clearly an important part of the subject, while the purpose of understanding the hardware and system appears to be orienting the user about the equipment.

3.5.2 Learning to Design Systems and Include Prefabricated Components

To design systems implies that pupils should be made aware of the systemic characteristics of any artefact they create, which are typically mechanical, electronic, and/or computer-based. The systemic characteristics become apparent when prefabricated components (such as motors or control units) are integrated into pupil's design. Thereby, the product's systemic character (consisting of several distinguishable parts that contribute to a common function) is highlighted.

The origins of the Finnish sloyd subject are in classic handicraft, although the subject has been modernized with the inclusion of more current materials and equipment. In the listed core content, there is a section about testing or trying out (Swedish 'pröva' can mean both 'try' and 'test'), which includes programming skills and the use of embedded systems in created products (Utbildningsstyrelsen, 2014, para. I3).

The English subject of design and technology has a strong design-and-make focus, with most of the core content being skills-oriented. The subject is described as what pupils should be able to do, not what they should know, describe, explain, or analyse. Moreover, the core content includes creating products in which 'more advanced electrical and electronic systems' or 'more advanced mechanical systems [that] enable changes in movement and force' are included (Department for Education, 2013, p. 3).

3.5.3 Learning to Model Systems and Use System Models

System modelling is the act of creating system models (i.e. simplified representations of systems). For example, models can reveal a system's physical or logical structure or be used to predict (or even explain) its behaviour under certain circumstances. When system modelling or the use of system models is described in the curricula and standards, it almost exclusively concerns systems' technical cores. The creation of models of artefacts is mentioned in several of the studied curricula and standards, mainly in connection with design-and-make activities. In the Swedish syllabus, it is mentioned that pupils should work with digital and physical models (Skolverket, 2018, p. 299). In Wales, pupils use three-dimensional models to document and communicate their ideas and intentions (Department for Children, Education, Lifelong Learning and Skills, 2008, p. 6; 2009). However, physical models that are explicitly intended to represent systems and highlight their systemic characteristics are missing from the studied curricula and standards.

Learning to create and use symbolic and mathematical models of systems is described as a learning objective in some curricula. This appears to be especially prevalent in curricula and standards where technology is described together with (or closely related to) the natural sciences. According to the categorisation of Nordlöf et al. (2021), this equates to a curriculum focusing on *technological scientific knowledge*.

A good example is the Danish curriculum for technology and natural science, which is divided into different competence areas (including modelling). Explicitly, it states that there should be a strong focus on pupils creating models and using them for problem-solving (Børne- og undervisningsministeriet, 2019, p. 44). This competence area is practised in relation to the curriculum's different themes, including technology and resources, the human body, the environment, and weather. In the technology and resources theme, teaching should be based on the pupils' use of process models for electricity and water systems in Year 6. Concurrently, they should also learn to use models in physics, chemistry, and geography (p. 76).

In Massachusetts' standards for science, technology, and engineering education, creating symbolic models of technical systems is practised repeatedly and with increasing levels of complexity. In Year 5, pupils use drawings to show relationships between parts of a device. In Years 9–12 they develop and use models based on evidence (including mathematical and computational) to predict phenomena. They are also encouraged to revise and refine their models based on observations. These models are of systems within the whole range of subjects that are covered by the standards, not just technological systems (Massachusetts Department of Elementary & Secondary Education, 2016, pp. 51, 110–111).

3.5.3.1 Learning to Analyse Systems' Technical Cores

The technical core of a system comprises its physical components, which is an aggregate of technical artefacts (Ewertsson & Ingelstam, 2004; Hughes, 1983; Kroes et al., 2006). In this case, analysis means using system models (which may be mental representations) to draw conclusions about the actual system and its behaviour. A thorough analysis also entails discussing and evaluating the validity and limitations of system models, often including explicit references to mathematics and/or natural sciences (see also Børne- og undervisningsministeriet, 2019, p. 14; de Kleer & Brown, 1981).

In curricula based mainly on *technical skills*, design and use of small mechanical and/or electrical systems are commonly included (see above). Even though it is not stated explicitly, it appears reasonable that the intention is that pupils should learn about similar systems through these exercises and be able to transfer this knowledge to non-educational contexts. Some texts also highlight the ability to analyse a system's technical core, with notable examples including the Australian and Danish curricula.

In the Australian curriculum, pupils' increased knowledge about (and ability to analyse) mechanical systems is described in the overview of the 'Engineering principles and systems' strand. In Years F–2 they '[e]xplore how technologies use forces to create movement in products'. Further, in Years 3–4, they 'investigate' said systems. Electricity is introduced in Years 5–6, and they use their knowledge to design 'simple, engineered solutions' including electromechanical systems in Years 7–8 (ACARA, 2015, p. 1).

The Danish curriculum recognizes pupil progress as model creators and model users. The models range from concrete models of everyday objects in Year 2, to models of electrical circuits and domestic technical systems in Year 6, and to wide-ranging supply or energy systems in Year 9. Further, it is stated that pupils should have learned to distinguish between model and reality by Year 2, and by Year 9 they should be able to discuss the usefulness and limitations of system models (Børne- og undervisningsministeriet, 2019, summarized on p. 18, progress in modelling competence on pp. 32–33). It should be noted that the acts and intentions of humans as parts of systems are not mentioned as parts of technological system models. Although the demands, wishes, and conflicts of interest within society concerning technical and industrial developments are mentioned in the curriculum (e.g. p. 14), no connection with system modelling or model use is made. The system modelling process of the Danish curriculum only appears to concern the system's technical core, relying on mathematics and natural sciences and not considering the social sciences.

3.5.3.2 Control Systems

When control systems are included in curricula, they are generally considered parts of systems' technical cores (see above). The reason for assigning them a separate heading is that in many cases they are treated separately from other components or sub-systems. They also tend to be described in abstract terms (such as input, output, control system). This is in stark contrast to other electrical and electronic equipment which is described mainly using terms referring to physical objects such as 'motor', 'diode', 'transistor', and 'integrated circuit' (e.g. National Council for Curriculum and Assessment, 2006). Control technologies and control theory are mentioned very briefly. Control equipment is described as 'black boxes', with 'feedback' and 'sensor' as the only field-specific terms that are commonly used. That the 'black boxes' contain transistors is not mentioned. The control equipment remains abstract.

3.5.4 Learning About Infrastructure and/or Industrial Manufacturing Systems

Infrastructure and industrial manufacturing systems cannot be brought into the classroom and generally cannot be handled, manipulated, or used by pupils. Studying these systems is mainly accomplished through the use of various models (or simulations), or study visits where the system or parts of it can be studied in real life.

In the Finnish curriculum, it is stated that pupils should study industry virtually or in actuality (Utbildningsstyrelsen, 2014, para. I7). However, the main purpose is to study the role of practical skills in working life, not the systemic aspects. In Denmark, pupils are expected to model and study electricity and water systems (see above). In Sweden, infrastructure has quite a prominent position in the technology syllabus. It should be introduced in Years 4–6 in the form of '[c]ommon technical systems at home and in society, such as networks for data communication, water and sewage systems', and in Years 7–9 through the 'internet and other global technical systems. The benefits, risks and limitations of these systems' (Skolverket, 2018, pp. 298, 299). However, detailed guidance pertaining to how pupils study these systems is not described in the curriculum. Although listed under the heading 'Technology, man, society and the environment', whether pupils should describe systems, use system models, create system models, or analyse the function of systems remains in control of the teacher.

3.5.5 *Learning to Analyse Socio-Technical Systems*

Socio-technical understanding pertains to the roles of technological systems in society (mainly large systems such as energy, transportation, and the Internet), currently, throughout history, and in the future. This category is different because it is a perspective rather than a skill or ability. The reason for according it a separate category (not within variations of *analysis* or *modelling*) is how it is included in the syllabi. In New Zealand (Ministry of Education, 2018), it has its separate 'strand', entitled 'Nature of technology'. In Sweden, it has its own heading in the syllabus: 'Technology, man, society and the environment' (Skolverket, 2018). Other subject content (such as the more design-and-make or engineering related) tends to be grouped according to skills, whereas the social aspects of technology are separated according to content.

3.5.6 *Developing Systems Thinking*

In the studied documents, the expression 'systems thinking' is used by ITEA (2007) and the National Research Council (2010, 2012). All three documents are from the United States, all were published approximately a decade ago, and none of them constitutes a curriculum or syllabus. This does not mean that the complex, ambiguous and rather vague concept of systems thinking is missing from the other documents; it simply means that the actual expression is not used. Predominantly, systems thinking can be regarded as a combination of the above categories, especially those that address analysis and modelling (see also ACARA, n.d., ITEA, 2007; Rosenkränzer et al., 2017). Systems thinking features strongly in the Danish curriculum (with its emphasis on modelling and systems of various types and sizes) and the Massachusetts curriculum. These curricula describe systems and modelling as general concepts and compare the features of technological systems and natural systems.

3.6 Discussion and Conclusion

Today's technologies are systemic in nature. Even those that we tend to regard as solitary artefacts contain multiple components that interact. Further, these artefacts often depend on external systems for functioning, such as the electrical supply system. Moreover, many artefacts become parts of larger systems when in use. For example, cars become parts of the road traffic system, and computers become parts of the Internet. Accordingly, understanding systems as a whole and as aggregates of components are essential aspects of pupils' understanding of today's technological world. This understanding is important for informed citizens in general, and for future engineers, technicians, artisans, technology teachers, and other professionals in the vast technological domain.

The purpose of this chapter is to provide an overview of how (and to what extent) systems and systems thinking are parts of national curricula and standards in technology education. A limited selection of documents was taken into consideration, which addresses technology (engineering) education for children and adolescents in primary and lower secondary school. The documents describe elementary technology education for future technology professionals as well as for most everyone else. The documents differ in style, level of detail, and content. For example, some of them describe technology in a quite narrow sense while others include natural science subjects. Therefore, comparing or evaluating these differing curricula becomes difficult and any attempt to adopt this approach would probably not generate any interesting results. In the documents, system-related content is more common than the term 'system', and systems thinking is far more common than the actual term 'systems thinking'. It could be construed that systems hide among the words describing components, products, processes, or interacting parts.

Pupils should learn about various aspects of systems by using them, creating models of them, and studying them in different ways. The almost total lack of human content in relation to systems is striking when reading the documents. When society or individual human beings are mentioned, it is because they affect or are affected by the system (e.g. Skolverket, 2018, p. 297). Humans are not mentioned as being parts of systems. This is probably a result of technology subjects being based on crafts and engineering rather than the social sciences. In the words of Nordlöf et al. (2021), they are dominated by *technical skills* and *scientific technological knowledge* rather than *socio-ethical technical understanding*. Further, including the roles of human agents in system models and system studies could add new perspectives, as the system is rendered both more flexible and less predictable. This could provide learning opportunities concerning work environments and ergonomics, which are included in a few curricula (e.g. Department of Basic Education, 2011; Utbildningsstyrelsen, 2014).

There are common features in most types of system studies, such as identifying how individual components contribute to the function of the whole system. However, we cannot pretend that system studies in introductory technology education is a unified whole with a definitive meaning. With a few exceptions, systems are not identified with their own theme or strand in the curricula, but they are mentioned in relation to other technical phenomena. The curricula and standards based on designing and making (e.g. England, Finland, Scotland, Wales) emphasize small systems that students can design or include in their designs. By comparison, curricula and standards that also include social aspects of technology (e.g. Australia and Sweden) include larger systems and some of their social aspects. In addition, curricula and standards that cover technology together with the natural sciences (e.g. Denmark and Massachusetts) describe cross-disciplinary system concepts that can supposedly be applied in both technological and other contexts. Finally, although teaching and learning about technological systems can (and should) incorporate many different facets, they always relate to collections of components functioning together.

References

[ACARA] Australian Curriculum Assessment and Reporting Authority. (2015). *Design and Technology: Sequence of Content F–10.* Retrieved from the ACARA website, https://docs.acara.edu.au/resources/Design_and_Technologies_-_Sequence_of_content.pdf

[ACARA] Australian Curriculum Assessment and Reporting Authority. (n.d.). *The Australian Curriculum: Learning areas: Technologies; Subjects: Design and Technologies, Digital Technologies.* Retrieved from the ACARA website.

Børne- og undervisningsministeriet [Danish Ministry of Children and Education] (2019). *Natur/teknologi. Faghaefte 2019.* Retrieved from EMU the Danish national portal for education, https://emu.dk/sites/default/files/2020-09/Gsk_fagh%C3%A6fte_naturteknologi.pdf

Braun, V., & Clarke, V. (2006). Using thematic analysis in psychology. *Qualitative Research in Psychology, 3*(2), 77–101.

de Kleer, J., & Brown, J. S. (1981). Mental Models of Physical Mechanisms and Their Acquisition. In J. R. Anderson (Ed.), *Cognitive Skills and Their Acquisition* (pp. 285–310). Lawrence Erlbaum Associates, Publishers.

Department for Children, Education, Lifelong Learning and Skills [in Wales]. (2008). *Design and Technology in the National Curriculum for Wales. Key Stages 2–3.* Cardiff, United Kingdom: Welsh Assembly Government. Retrieved from https://hwb.gov.wales/curriculum-for-wales-2008/key-stages-2-to-4/design-and-technology-in-the-national-curriculum-for-wales/

Department for Children, Education, Lifelong Learning and Skills [in Wales]. (2009). *Design and Technology. Guidance for the Key Stages 2 and 3.* Cardiff, United Kingdom: Welsh Assembly Government. Retrieved from https://hwb.gov.wales/curriculum-for-wales-2008/key-stages-2-to-4/design-and-technology-in-the-national-curriculum-for-wales-guidance-for-key-stages-2-and-3/

Department for Education [in England] (2013). *Design and Technology Programmes of Study: Key Stage 3.* Retrieved from https://assets.publishing.service.gov.uk/government/uploads/system/uploads/attachment_data/file/239089/SECONDARY_national_curriculum_-_Design_and_technology.pdf

Department of Basic Education [of South Africa]. (2011). *Curriculum and Assessment Policy Statement: Technology.* Pretoria, South Africa: Department of Basic Education. Retrieved from https://www.education.gov.za/Portals/0/CD/National%20Curriculum%20Statements%20and%20Vocational/CAPS%20SP%20%20TECHNOLOGY%20GR%207-9.pdf?ver=2015-01-27-160218-077 on 30 January 2021.

Ewertsson, L., & Ingelstam, L. (2004). Large Technical Systems: A Multidisciplinary Research Tradition. In M.-O. Olsson & G. Sjöstedt (Eds.), *Systems Approaches and their Applications* (pp. 291–309). Kluwer Academic Publishers.

Frank, M. (2002). What is 'engineering systems thinking'? *Kybernetes, 31*(9–10), 1350–1360.

Hughes, T. P. (1983). *Networks of Power. Electrification in Western Society, 1880–1930.* Baltimore, MD: John Hopkins University Press.

[IEEE] International Electrical and Electronics Engineers. (2007). Systems engineering—Application and management of the systems engineering process. ISO/IEC 26702, IEEE Std 1220–2005. New York: IEEE Computer Society. Retrieved from https://ieeexplore.ieee.org/servlet/opac?punumber=4278474

[ITEA] International Technology Education Association [today known as ITEEA, International Technology and Engineering Educators Association]. (2007). *Standards for Technological Literacy.* Reston, VA: International Technology Education Association.

Krippendorff, K. (2013). *Content analysis: An introduction to its methodology.* Sage Publications.

Kroes, P., Franssen, M., van de Poel, I., & Ottens, M. (2006). Treating Socio-technical systems as engineering systems: some conceptual problems. *Systems Research and Behavioral Science, 23,* 803–814.

Magee, C. L. & de Weck, O. L. (2002). *An Attempt at Complex System Classification.* ESD Internal Symposium, Working Paper Series ESD-WP-2003–01.02. Cambridge, MA: Massachusetts Institute of Technology, Engineering Systems Division. Retrieved from https://dspace.mit.edu/bitstream/handle/1721.1/102730/esd-wp-2003-01.02.pdf?sequence=1

Massachusetts Department of Elementary and Secondary Education (2016). *2016 Massachusetts Science and Technology/Engineering Curriculum Framework.* Retrieved from the Massachusetts Department of Elementary and Secondary Education website: http://www.doe.mass.edu/frameworks/scitech/2016-04.pdf

Ministry of Education [in New Zealand] (2018). *Technology in the New Zealand Curriculum.* Retrieved from the Ministry of Education's website: https://nzcurriculum.tki.org.nz/content/download/168478/1244184/file/NZC-Technology%20in%20the%20New%20Zealand%20Curriculum-Insert%20Web.pdf

National Academy of Engineering. (2010). *Standards for K–12 Engineering Education?* The National Academies Press.

National Council for Curriculum and Assessment [in Ireland] (2006). *Leaving Certificate. Technology Syllabus.* Retrieved from the National Council for Curriculum and Assessment website: https://www.curriculumonline.ie/getmedia/da63d79a-b84f-4b8c-ba06-a4cb1fd21e07/SCSEC34_Technology_syllabus_eng.pdf

National Research Council [in the United States of America]. (2010). *Standards for K–12 Engineering Education?* Washington, DC: The National Academies Press. Retrieved from https://www.nap.edu/catalog/12990/standards-for-k-12-engineering-education

National Research Council [in the United States of America]. (2012). *A Framework for K–12 Science Education. Practices, Crosscutting Concepts, and Core Ideas.* Washington, DC: The National Academies Press. Retrieved from http://www.nap.edu/catalog.php?record_id=13165

National Research Council [in the United States of America]. (2013). *Next Generation Science Standards: For States, By States.* Washington, DC: The National Academies Press. Retrieved from https://www.nap.edu/catalog/18290/next-generation-science-standards-for-states-by-states

Nordlöf, C., Norström, P., Höst, G., & Hallström, J. (2021). Towards a Three-part Heuristic Framework for Technology Education. *International Journal of Technology and Design Education.* Advance online publication. https://doi.org/10.1007/s10798-021-09664-8

Norström, P. (2016). The Nature of Pre-University Engineering Education. In M. J. de Vries, L. Gumaelius, & I.-B. Skogh (Eds.), *Pre-University Engineering Education* (pp. 27–46). Sense Publishers.

Olsson, M.-O., & Sjöstedt, G. (2004). Systems and System Theory. In M.-O. Olsson & G. Sjöstedt (Eds.), *Systems Approaches and their Applications* (pp. 3–29). Kluwer Academic Publishers.

Rosenkränzer, F., Hörsch, C., Schuler, S., & Riess, W. (2017). Student teachers' pedagogical content knowledge for teaching systems thinking: Effects of different interventions. *International Journal of Science Education, 39*(14), 1932–1951. https://doi.org/10.1080/09500693.2017.1362603

Schuler, S., Fanta, D., Rosenkraenzer, F., & Riess, W. (2018). Systems thinking within the scope of sustainable development (ESD)—A heuristic competence model as a basis for (science) teacher education. *Journal of Geography in Higher Education, 42*(2), 192–204.

Scottish Qualifications Authority. (2012). *National 3 Design and Technology Course Specification (C720 73)*. Glasgow, United Kingdom: Scottish Qualification Authority. Retrieved from https://www.sqa.org.uk/files/nq/CfE_CourseSpec_N3_Technologies_DesignandTechnology.pdf

Skolverket [The Swedish National Agency for Education]. (2018). *Curriculum for the Compulsory School, Preschool Class and School-Age Ecucare*. Stockholm: Skolverket. Retrieved from https://www.skolverket.se/publikationer?id=3984

Utbildningsstyrelsen [The Finnish National Agency for Education]. (2014). *Grunderna för läroplanen för den grundläggande utbildningen 2014: Slöjd.* Retrieved from https://eperusteet.opintopolku.fi/#/sv/perusopetus/419550/vuosiluokkakokonaisuus/428782/oppiaine/530524

Per Norström is associate professor (Swedish: *universitetslektor*) of technology education at KTH Royal Institute of Technology, Stockholm, Sweden. His research interests mainly concern pre-university engineering education and the epistemology of technology.

Chapter 4
Cross-Curriculum System Concepts and Models

Maria Svensson

Abstract Systems thinking is a tool for understanding that one could learn to use gradually through practice and continual improvement in relation to different subject areas. In the subject technology, as well as biology, systems are part of the curriculum but there is a lack of knowledge about cross-curricular learning about systems. In this chapter, a qualitative system literature review is used to present and compare system concepts and models used in two different subjects, biology and technology. Furthermore, a reflection on what system aspects might contribute to cross-curricular learning opportunities is done using the structure, behaviour and function (SBF) system thinking model. In the analysis of the 22 articles, 12 about biology education and 10 about technology education, similarities and differences in the structural and behavioural aspects between the two subjects stand out. On the other hand, it became clear that the functional aspect only occurs in relation technological systems. There are system aspects that cross over fields that might have potential for new ways of teaching about systems and develop systems thinking. This is not least important for developing understanding and preparedness to address sustainability issues today and tomorrow, something that all teachers have a responsibility to do.

Keywords Systems thinking · Structure · Behaviour · Function (SBF) model · Cross-curricular · Variation theory · Literature review

Pupils in schools are often faced with many different subjects during one day. They have a biology lesson on oxygen function in the body in the morning, and after lunch, they go to a lesson in history about treadmills and their significance for industrial development. Teachers in the different subjects often do not know, or think about, the similarities between their subjects, they rather indicate the differences that exist between one's own subject and other subjects. However, there are concepts and models that might enhance pupils' learning if they were used in various subjects.

M. Svensson (✉)
Department of Pedagogical, Curricular and Professional Studies, University of Gothenburg, Gothenburg, Sweden
e-mail: maria.svensson@ped.gu.se

J. Hallström and P. J. Williams (eds.), *Teaching and Learning about Technological Systems*, Contemporary Issues in Technology Education,
https://doi.org/10.1007/978-981-16-7719-9_4

The aim of this chapter is to present and compare system concepts and models used in two different subjects, biology and technology, and reflect on how an awareness regarding system aspects may contribute to cross-curricular learning opportunities. Learning about systems should include, amongst other characteristics, a focus on methodologies that will foster collaboration, discussion, and reflection (Jacobson & Wilensky, 2006). However, a unified conceptual framework for the development of systems thinking in education is still absent from most schools' curricula (Jacobson & Wilensky, 2006; Plate, 2010).

When I started my research about technological systems almost 20 years ago, I had been teaching technology and science for some years both in teacher education and in compulsory school. During these years, I had on several occasions been struck by the untapped opportunities that exist when it comes to systems thinking. Systems thinking is a tool for understanding that one could learn to use gradually through practice and continual improvement in relation to different subject areas (Mella, 2012). When I was recently observing teaching in biology and technology for student teachers, it came to my consciousness again that systems thinking could be more effectively used in teaching. One session I observed was about the human body, and it was clear that students encounter systems in various subjects without us, as teacher educators, using this opportunity to develop an understanding of systems and systems thinking as a model to see both the parts and the whole of a phenomenon. Seng (2006) proposed that systems thinking is an effective approach for observing reality and constructing sensible and coherent models which make us look for connections in the world around us. As teachers and teacher educators, we could make better use of systems thinking to describe and understand phenomena, to encourage our students and student teachers to see the connections and understand the whole in relation to different subject areas. In the subjects, biology and technology, there are several areas where systems are present. Therefore, in this chapter, I focus on these two subjects when striving for a more developed conceptual framework of systems and systems thinking in education. A literature review of system concepts and models in biology and technology education forms the basis for the comparisons and further reasoning regarding the development of system concepts and models in different subjects.

4.1 Background

Systems thinking can be described as an ability to recognize, describe and model complex aspects of reality as systems. This implies identifying important elements of the system and the varied interdependency between these elements. Mella (2012) describes systems thinking by five rules, where the first one obliges us to "see the trees and the forest" (p. 9).

> To understand reality, we must not limit ourselves to observing only individual objects, elements, or entities; it is necessary to "see" even larger groupings that these compose, attributing to them an autonomous meaning. The converse process is also true: we cannot

limit our selves only to considering an object in its unity but must force ourselves "to see" its component parts (Mella, 2012, p. 9).

Systems thinking is a holistic approach for examining complex problems and systems that focuses on the interactions amongst system components and the patterns that emerge from those interactions. Systems thinking can help students develop higher-order thinking skills in order to understand and address complex, interdisciplinary, real-world problems (Assaraf & Orion, 2005). Because of these potential benefits, there have been efforts to support the implementation of systems thinking approaches as a cross-curricular method (Forrester, 1993; Sweeney, 2005). Even though there are differences between the systems which are evident in biology and technology, there is potential for the development of general understandings of systems thinking that could support higher-order thinking if effectively promoted in education. Ho (2019) stated that to help students be better equipped to solve problems involving complex systems, it is important to find ways to incorporate systems and systems thinking in education, and in that way enable students to analyse and understand system characteristics. Systems thinking approaches in education are increasingly widespread in disciplines such as biology, engineering, geoscience and sustainable development, but there is yet more to learn about how to develop cross-curricular concepts relevant to systems and systems thinking. There is a lack of research investigating cross-curricular system concepts and models within education.

Making sense of complex systems requires that a person constructs a network of concepts and principles about some domain that represents key phenomena and the interrelationships amongst different levels of the system, whether it is macro to micro or structure to function (Goel et al., 2009; Hmelo-Silver & Pfeffer, 2004). Research has demonstrated that people can transfer deep principles of complex systems across domains when examined in the context of simulations (Goldstone & Sakamoto, 2003). Thus, to be able to transfer knowledge from one context to another it is essential to discern both similarities and differences between contexts (Marton, 2006). Possible aspects to discern in one situation might be irrelevant to another situation. However, to be able to transfer knowledge it is essential to see different things of the same sort (Marton, 2014). In relation to systems components and the connections between components are used to describe a system, in that way they can be understood as the same, interrelated components. However, the nature and function of the components differ depending on the context in which the system exists. In a biological system such as an ecosystem, components are used to describe a food web with animals and green plants. In a technological system, such as a wastewater system, components are used to describe pipes and pumps. Even though these components are different when it comes to their properties and functions in each system, they are understood as components in a system, parts that make up the whole. When teaching about systems in different subjects, one should be aware of aspects that could be experienced as different but at the same time be examples of the same. Components in an ecosystem are different from the ones in a wastewater system but they all play a role in the system as components, this means seeing them as similar in one sense and different in another. Using and understanding this in teaching can help students prepare for the unknown by the means of the known.

4.2 Method

Several models have been developed as conceptual representations of systems thinking which make different levels explicit. The literature review presented in this chapter builds on a sample of research about systems from 2010–2020 using an approach described as structure–behaviour–function (SBF) thinking (see also chapter by Mioduser) (Goel et al., 2009). Hmelo-Silver and Pfeffer (2004) suggest that structure–behaviour–function (SBF) theory may provide a structure for thinking about complex systems in different areas. In the SBF thinking model, the different levels of a system, in terms of structures, behaviours and functions, and their interconnections can be identified (Goel et al., 2009; Hmelo-Silver & Pfeffer, 2004; Hmelo-Silver et al., 2008). The *structure* refers to parts of the system that vary in size and organization and answer the "what" of the system, meaning the components of the system as well as the connections between them. The *behaviour* specifies the "how" of the system, the processes occurring in the system, and the *function* refers to the role or output of the system or subsystems concerning the "why" question, the purpose of the system. The SBF system thinking model has been used for explaining and justifying the design of physical devices such as electrical circuits and heat exchangers (Goel et al., 1996) as well as the respiratory system and an aquarium ecosystem (Hmelo et al., 2000; Hmelo-Silver & Pfeffer, 2004).

4.2.1 *The Literature Review*

A literature review is useful when the aim is to provide an overview of a certain issue such as system concepts and models in biology and technology education. "Typically, this type of literature review is conducted to evaluate the state of knowledge on a particular topic" (Snyder, 2019, p. 334). To find the characteristics of systems in the subjects biology and technology, a literature review of academic journal papers published from 2010 to 2020 was conducted. Sources were limited to peer-reviewed academic journal papers that are indexed, reliable and searchable because these have a rigorous publication procedure. Therefore, the quality of the journal papers can be trusted (Snyder, 2019). A qualitative systematic review is used to interpret and broaden the understanding of a particular phenomenon, here system concepts and models in biology and technology education (Grant & Booth, 2009). By using a literature review, an identification of what has been accomplished within the area is possible and allows for comparisons of findings from qualitative studies in the two subject areas.

The databases Educational Resources Information Centre (ERIC) and ProQuest were used as a source to find articles that could give a relevant base to present and compare system concepts and models used in two different subjects, biology and technology. Keywords that were used to find relevant journal papers were in the first step a combination of *biology education* and *system*, or *technology education* and *system* within the abstract of a full text peer-reviewed paper between 2010 and 2020.

Biology education AND system - 116 hits

Technology education AND system – 1975 hits

The next step was to narrow the search of technology-related papers using the keywords *technology education* and *technological system*. Technology education is in many countries only connected with computer science, and in this study, technology is understood in a broader perspective where technology refers to human-made artefacts and systems that solve a problem or fulfil a desire (Mitcham, 1994). Therefore, including *technological* in combination with *system* addresses systems of complex, problem-solving components that solve problems or fulfil goals using available means and directed to different kinds of technological areas (Huges, 1987). This was thus a way of trying to reduce the number of papers that solely focus on technology as computer science or as other digital tools.

Technology education AND technological system – 206 hits

Systems are often connected to computers, described as systems per se or as parts of larger systems. The purpose of this study is to focus on the subjects technology and biology in combination with system, not with the main focus on the use of computers as digital tools in teaching. Therefore, in a try to narrowing the search further in both groups', the word computer*, was excluded from the list of 206 hits and 116 hits.

Technology education AND technological systems NOT computer* - 74 hits

Biology education AND systems NOT computer* - 50 hits

After this selection within the database, the abstracts were read carefully to identify those relevant to the school subjects biology and technology in combination with system concepts and models. In this step of the analysis, exclusion criteria were education systems, management systems or the use of technology as pedagogical tools rather than system aspects of a subject. This analysis ended up with 10 technology education papers and 12 biology education papers listed in Tables 4.1 and 4.2.

4.2.2 *The SBF Systems Thinking Model*

To analyse the remaining papers and break down, the content in relation to system concepts and models the SBF system thinking model (Goel et al., 2009; Hmelo-Silver & Pfeffer, 2004) was used. The purpose was also to identify similarities and

Table 4.1 Technology-related papers presented in alphabetic order of the first author

Technology–10 papers			
Author	Title	Journal	Year
Autio, Ossi; Olafsson, Brynjar; Thorsteinsson, Gisli	Examining Technological Knowledge and Reasoning in Icelandic and Finnish Comprehensive Schools	Design and Technology Education Vol. 21, Iss. 2,	2016
Compton, Vicki J; Compton, Ange D.	Teaching Technological Knowledge: Determining and Supporting Student Learning of Technological Concepts	International Journal of Technology and Design Education Vol. 23, Iss. 3,	2013
Hallström, Jonas; Klasander, Claes	Visible Parts, Invisible Whole: Swedish Technology Student Teachers' Conceptions about Technological Systems	International Journal of Technology and Design Education Vol. 27, Iss. 3,	2017
Harsh, Matthew; Bernstein, Michael J.; Wetmore, Jameson; Cozzens, Susan; Woodson, Thomas; et al	Preparing Engineers for the Challenges of Community Engagement	European Journal of Engineering Education Vol. 42, Iss. 6,	2017
Hope, Gill	Designing Technology: An Exploration of the Relationship between Technological Literacy and Design Capability	Design and Technology Education Vol. 18, Iss. 2,	2013
Jung, Kiho; Otaka, Yuki	The Introduction of a Thin-Bending Wood Horn Speaker as Multipurpose Teaching Material in Japanese Junior High School Technology Classes	World Journal of Education Vol. 9, Iss. 6,	2019
Park, Wonyong	Beyond the 'Two Cultures' in the Teaching of Disaster: Or How Disaster Education and Science Education Could Benefit Each Other	Educational Philosophy and Theory Vol. 52, Iss. 13,	2020
Schooner, Patrick; Nordlöf, Charlotta; Klasander, Claes; Hallström, Jonas	Design, System, Value: The Role of Problem-Solving and Critical Thinking Capabilities in Technology Education, as Perceived by Teachers	Design and Technology Education Vol. 22, Iss. 3,	2017
Schooner, Patrick; Klasander, Claes; Hallström, Jonas	Swedish Technology Teachers' Views on Assessing Student Understandings of Technological Systems	International Journal of Technology and Design Education Vol. 28, Iss. 1,	2018

(continued)

Table 4.1 (continued)

Technology–10 papers			
Author	Title	Journal	Year
Svensson, Maria; Ingerman, Ake	Discerning Technological Systems Related to Everyday Objects: Mapping the Variation in Pupils' Experience	International Journal of Technology and Design Education Vol. 20, Iss. 3,	2010

differences in system levels between the papers in biology and technology and reflect on how an awareness regarding system aspects in education may contribute to cross-curricular learning opportunities. The articles were read, the purpose of the articles was identified and the system aspect that was the focus of the articles identified using the SBF systems thinking model where the articles focusing mainly on the organization, size and components of a system were categorized as a structural system aspect. Articles that on the other hand focus on the processes in the system and how the system works were categorized as behavioural system aspects. If the article had a strong connection to the purpose of the system and answers the why-question, they were categorized with a focus on functional system aspects.

4.2.3 Limitations of the Study

A common problem when using a literature review is the risk of making limitation of samples too narrow and failing to describe in detail how the literature review was conducted (Snyder, 2019). With this in mind, the different steps in the selection of relevant articles are described and the process of limiting the search is motivated, but nevertheless, it is possible that articles of relevance fall outside this search because of the chosen limitations. It is also the case that the database used does not cover all articles within the field even if it is a database used for articles relevant in education, which becomes another source of error. Despite this, it is possible to describe indications on system concepts and models used in biology and technology, and reflect on how an awareness regarding system aspects in education may contribute to cross-curricular learning opportunities.

4.3 Results

The SBF systems thinking model, applied to the identified literature about systems in technology and biology education, makes differences and similarities between systems discernible (see Table 4.3). The results of how systems are described in the two subjects are presented first and after that the identified similarities and

Table 4.2 Biology-related papers presented in alphabetic order of the first author

Biology–12 papers			
Author	Title	Journal	Year
Akçay, Süleyman	Prospective Elementary Science Teachers' Understanding of Photosynthesis and Cellular Respiration in the Context of Multiple Biological Levels as Nested Systems	Journal of Biological Education Vol. 51, Iss. 1,	2017
Ballen, J. Cissy & Greene, W. Harry	Walking and talking the tree of life: Why and how to teach about biodiversity	PLOS Biology 15(3)	2017
Berat, AHİ	Thinking about digestive system in early childhood: A comparative study about biological knowledge	Cogent Education; Abingdon Vol. 4, Iss. 1,	2017
Boersma, Kerst; Waarlo, Arend Jan; Klaassen, Kees	The Feasibility of Systems Thinking in Biology Education	Journal of Biological Education Vol. 45, Iss. 4,	2011
Çuçin, Arzu; Özgür, Sami; Güngör Cabbar, Burcu	Comparison of Misconceptions about Human Digestive System of Turkish, Albanian and Bosnian 12th Grade High School Students	World Journal of Education Vol. 10, Iss. 3,	2020
Dam, Michiel; Ottenhof, Koen; Carla Van Boxtel; Janssen, Fred	Understanding Cellular Respiration through Simulation Using Lego® as a Concrete Dynamic Model	Education Sciences; Basel Vol. 9, Iss. 2,	2019
Hart, Emily R.; Webb, James B.; Danylchuk, Andy J.	Implementation of Aquaponics in Education: An Assessment of Challenges and Solutions	Science Education International Vol. 24, Iss. 4,	2013
Kattmann, Ulrich	A Biologist's Musing on Teaching about Entropy and Energy: Towards a Better Understanding of Life Processes	School Science Review Vol. 99, Iss. 368,	2018
Knippels, Marie-Christine P. J.; Arend, Jan Waarlo	Development, Uptake, and Wider Applicability of the Yo-yo Strategy in Biology Education Research: A Reappraisal	Education Sciences, Vol. 8, Iss. 3,	2018

(continued)

Table 4.2 (continued)

Biology–12 papers			
Author	Title	Journal	Year
van Mil, Marc H. W.; Boerwinkel, Dirk Jan; Waarlo, Arend Jan	Modelling Molecular Mechanisms: A Framework of Scientific Reasoning to Construct Molecular-Level Explanations for Cellular Behaviour	Science & Education Vol. 22, Iss. 1,	2013
Ozgur, Sami	The Persistence of Misconceptions about the Human Blood Circulatory System amongst Students in Different Grade Levels	International Journal of Environmental and Science Education Vol. 8, Iss. 2,	2013
Tripto, Jaklin; Assaraf, Orit Ben; Snapir, Zohar; Amit, Miriam	How Is the Body's Systemic Nature Manifested amongst High School Biology Students?	Instructional Science: An International Journal of the Learning Sciences Vol. 45, Iss. 1,	2017

differences. In the discussion, these results are used to elaborate on how awareness regarding system aspects used in biology and technology education may contribute to cross-curricular learning opportunities.

The technology-related articles are mainly linked to functional aspects, and the biology-related articles are to a greater extent linked to behavioural aspects. The similarities are mainly related to structural aspects, although there are differences in the two subject areas in how one chooses to describe the structure, as levels and/or as components.

4.3.1 Systems in Biology Education

Research about biological systems in education is related to understanding concepts with a structural character, described as different levels in the systems and how these levels are connected (see, e.g., Knippels & Waarlo, 2018). It is about organizing the system with a focus on the size of different parts, for example starting on the level of the organism and descend from there to the level of the organ and the cell and to ascend to the level of the population and community (Boersma et al., 2011).

In relation to the behavioural aspect, flows of resources are described as a concept of energy and matter related to ecosystems and systems in the body (Akçay, 2017; Çuçin et al., 2020). There are also examples of studies that identify misconceptions in relation to learning about systems in the human body such as circulation and cell systems (see, e.g., Çuçin et al., 2020). The misconceptions are connected to the structural aspect, in the choice of components used to describe the system, as well as

Table 4.3 Overview of the literature and the SBF aspects

Subject area	Article	SBF aspects		
		Structure	Behaviour	Function
Biology education	Akçay, Süleyman		**Flow**—of energy and matter in photosynthesis and cellular respiration **Process**—interactions between energy and matter in ecosystems	
	Ballen, J. Cissy & Greene, W. Harry	**Level**—taxonomy to organizing details of biodiversity		
	Berat, AHİ	**Component**—part in the digestive system	**Process**—what happens in organs	
	Boersma, Kerst; Waarlo, Arend Jan; Klaassen, Kees	**Level**—starting in something small (an organism) and ending in something bigger (the population)		
	Çuçin, Arzu; Özgür, Sami; Güngör Cabbar, Burcu	**Component**—drawings of the human digestive systems with different components **Connection**—between different components in the human digestive systems	**Flow**—input and output	
	Dam, Michiel; Ottenhof, Koen; Carla Van Boxtel; Janssen, Fred	**Level**—cellular respiration from microscopic to macroscopic	**Process**—activities that connect different organization levels	
	Hart, Emily R; Webb, James B; Danylchuk, Andy J	**Component**—physical parts		

(continued)

Table 4.3 (continued)

Subject area	Article	SBF aspects		
		Structure	Behaviour	Function
	Kattmann, Ulrich		**Flow**—energy flow in life processes **Process**—exergonic and endergonic reactions	
	Knippels, Marie-Christine PJ; Arend, Jan Waarlo	**Level**—starting in something small (a molecule) and ending in something bigger (a population)		
	van Mil, Marc H. W; Boerwinkel, Dirk Jan; Waarlo, Arend Jan	**Level**—starting in something small (a molecule) and ending in something bigger (a cell)	**Process**—activities that connect different levels	
	Özgur, Sami		**Flow**—blood circulation	
	Tripto, Jaklin; Assaraf, Orit Ben; Snapir, Zohar; Amit, Miriam	**Level -** hierarchy levels of the organism	**Flow**—matter **Process**—transfer between cells and environment	
Technology education	Autio, Ossi; Olafsson, Brynjar; Thorsteinsson, Gisli		**Process**—the functions in different parts in a mechanical system	**Purpose**—solving problems
	Compton, Vicki J; Compton, Ange D	**Component**—identifying parts in different systems **Connections**—between components to enable a specific function		

(continued)

Table 4.3 (continued)

Subject area	Article	SBF aspects		
		Structure	Behaviour	Function
	Hallström, Jonas; Klasander, Claes	**Component**—visible and invisible parts of the mobile phone, elevator and electric grid	**Flow**—energy, matter and information	
	Harsh, Matthew; Bernstein, Michael J; Wetmore, Jameson; Cozzens, Susan; Woodson, Thomas; et al		**Process**—the influence of human actions in different functions of a system	**Purpose**—solving problems and changing the conditions for humans and for the environment
	Hope, Gill			**Purpose**—solving problems and developing new tools, artefacts and environments
	Jung, Kiho; Otaka, Yuki	**Component**—parts included in a wood horn speaker	**Process**—the functions in different parts in the speaker	
	Park, Wonyong		**Process**—identifying failures in processes	**Purpose**—solving problems and changing the conditions for humans

(continued)

Table 4.3 (continued)

Subject area	Article	SBF aspects		
		Structure	Behaviour	Function
	Schooner, Patrick; Nordlöf, Charlotta; Klasander, Claes; Hallström, Jonas	**Components**—identifying parts in different systems	**Flow**—input–output of information **Process**—identifying interruptions in the process, "Black-boxing"	
	Schooner, Patrick; Klasander, Claes; Hallström, Jonas	**Component**—identifying parts in different systems	**Flow**—causality between input and output **Process**—how the parts and the whole of the system operating	**Purpose**—solving problems and changing the conditions for humans and for the society
	Svensson, Maria; Ingerman, Ake	**Component**—identifying parts in different systems. Micro and macrolevel **Connections**—between components to enable a specific function	**Flow**—material, energy and information **Process**—transportation, transformation, control and store of flow in different components	**Purpose**—solving problems and changing the conditions for humans

to the behavioural aspect in the descriptions of flow and processes in different body components.

No functional aspects are evident in the biology-related articles in this study which indicate that there is no specific focus on the role or output of the system or subsystems concerning the "why" question, the purpose of the system.

4.3.2 Systems in Technology Education

Research about technological systems shows that concepts that seem to have significance for learning and understanding of systems are related to components within the structural aspect (see, e.g., Compton & Compton, 2013; Schooner et al., 2018). Components are often described on a general level as physical parts of systems. Thus, in some studies the components mentioned are connected to specific content such as the mobile phone, elevator and electric grid (Hallström & Klasander, 2017) or the wooden horn speaker (Jung & Otaka, 2019). Different characteristics of components are identified and discussed in some studies, for example in Hallström and Klasander (2017), components as visible or invisible are mentioned, and in Svensson and Ingerman (2010) a way of analyzing components on three levels, the level of the object themselves, the micro-level inside objects, the macro-level outside the objects is suggested.

The behavioural aspect in technological systems describes the concepts of flow of resources as energy, matter and information. The flow is normally expressed as the input and output in the system. The processes within components, how the parts and the whole of the system operates, are also included in the behavioural aspect. This has to do with the functioning of the components in the system when transporting, transforming, controlling or/and storing the flow of resources (see, e.g., Schooner et al., 2017; Svensson & Ingerman, 2010).

In relation to the functional aspect, answering the "why" question, solving problems for humans in society seems to be an important concept mentioned in some articles in relation to the purpose of technological systems. There is also a focus on how the functional aspect of systems involves solving problems and changing the conditions for humans over time (see, e.g., Autio, Olafsson & Thorsteinsson, 2016; Park, 2020).

4.3.3 Summarizing the Result of the Literature Review

The results of the literature review indicate that there are system concepts that are used in both biology and technology education. In relation to the SBF systems thinking model, structural and behavioural aspects are found in the biology as well as in technology literature. However, the functional aspect is only identified as an aspect in the technology-related literature. An interpretation of this may be that the conducted studies in technology during the last 10 years have mainly focused on establishing technology as a separate subject. This requires clear motivations and arguments about the purpose of various teaching elements such as technological systems. Another interpretation is that systems in technology to a greater extent are still developing and changing while systems in biology are more established and have been used in teaching for a long time; therefore, their purpose does not have to be discussed. Furthermore, function in biology is more difficult to explain and discuss than in technology. In technology, there is always a system builder behind the technological system whereas in biological systems there is no system builder, described by Dawkins (1996) as the "blind watchmaker". In biological systems "natural selection is the blind watchmaker, blind because it does not see ahead, does not plan consequences, has no purpose in view" (Dawkin, 1996, p. 21). This might also be a reason why the functional aspect does not appear in the articles connected to biological systems.

In respect to the two investigated subjects, the system concepts which therefore have potential for contributing to cross-curricular learning opportunities are aspects related to the structure - *components as part of the system, characteristics of components* and *structural level of components,* and to the behaviour *–the character of flow (energy, matter, information), the input and output, the processes within components and between components and levels.*

4.4 Discussion

Teachers' and teacher educators' knowledge about the similarities and differences in the use of system concepts and models provides an opportunity to see recurring patterns in systems thinking between subjects and disciplines. Such knowledge is essential to meet the global problems we face, not least in relation to our environment (see, e.g., Rosenkränzer et. al., 2016). Systems thinking represents one such pedagogical approach, in which a holistic framework empowers both teachers and students to recognize how fundamental concepts, taught in the classroom, can be used as tools to better address complex, multicomponent modern challenges (Ho, 2019). The identified concepts of the SBF system thinking model have potential as a pedagogical approach, where structural and behavioural aspects in the two investigated subjects biology and technology open up possibilities for identifying things that vary in the mentioned aspects. This could be a first step towards transfer of system knowledge and contribute to cross-curricular learning opportunities.

To be able to generalize and transfer understanding of one systems concept into new contexts, the learner needs to develop the capability to discern differences and similarities of system concepts and the aspects that are critical for understanding the system concepts (Marton, 2006). However, using an understanding of the system concept in one context is not always easily transferred to another context. Magntorn and Helldén (2007) found this in their study about transferring system knowledge from one ecosystem (a forest), studied in detail, to another ecosystem (a pond).

They found that the structural and behavioural levels according to the definition of Hmelo-Silver and Pfeffer (2004) are difficult for the students to transfer to new environments, even though the use of the SBF system thinking model could be a first step in identifying similarities and differences in system concepts supplemented with discerned critical aspects.

Recognizing the three SBF aspects, structure, behaviour and function, as dimensions that can vary between systems, opens up opportunities for cross-curricular learning about systems and systems thinking. If we want to use systems thinking as a pedagogical approach, a way forward could be to use the teaching situations where systems are present but might not be at the core of the content, to learn to discern critical aspects of systems. For this to happen, we have to encounter and experience certain necessary patterns of variation and invariance (Marton, 2014). To give an example, we can think of students that have a lesson about the digestive system in the body, describing different parts of that system and how the food is processed in these parts. Using SBF we can see that the structure and the behaviour of the system are relevant system aspects in relation to this. The teacher, when teaching about this, could draw students' attention to the stomach as a component with processes for transforming matter (food) and compare it with components that have a similar purpose such as a cell or an engine. Identifying similarities (the transformation of matter) and differences (the properties of the structure) could be a way of developing students' awareness of aspects that vary in different system contexts. When looking at systems in biology and technology, it is evident that similarities and differences connected to structural and behavioural aspects are possible to use to visualize variation. In the current literature review, tentative critical aspects in relation to the concept structure and behaviour have been discerned and in that way offer a starting point for cross-curricular opportunities. On the other hand, problems arise with regard to the functional aspect if one tries to answer the question why does this system exist; there is no answer for a biological system but there is always a purpose with the technological systems conceived by man (Dawkins, 1996).

The aim of this chapter is to present and compare system concepts and models used in two different subjects, biology and technology and reflect on how an awareness regarding system aspects may contribute to cross-curricular learning opportunities. A question to ask is what do teachers gain from knowing about cross-curriculum system concepts and models? An answer to that after this literature review is that there are system aspects that cross over fields that might have potential for new ways of teaching about systems. The awareness of critical aspects of system concepts in one system context, for example, a biological system, enhances the likelihood of being able to discern the same and other critical aspects in another system context, for example, a technological system. To be aware of different aspects of systems concepts opens up an opportunity to understand what a system is in new and more powerful ways. It also enables a more nuanced way of understanding specific systems such as ecosystems or wastewater systems. Being aware of systems concepts and systems thinking approaches that exist today in different school subjects and being open to learning in a cross-curricular manner allow learners to better understand and manage various situations in their environment. This is crucial for being able to understand

and deal with sustainability problems in society, something that all teachers have a responsibility to prepare their students for, regardless of which school subject they teach.

References

Akçay, S. (2017). Prospective elementary science teachers' understanding of photosynthesis and cellular respiration in the context of multiple biological levels as nested systems. *Journal of Biological Education, 51*(1), 52–65.

Assaraf, O. B. Z., & Orion, N. (2005). A study of junior high students' perceptions of the water cycle. *Journal of Geoscience Education, 53*(4), 366–373.

Autio, O., Olafsson, B. & Thorsteinsson, G. (2016). Examining technological knowledge and reasoning in Icelandic and Finnish comprehensive schools. *Design and Technology Education,* 21(2).

Ballen, C. J., & Greene, H. W. (2017). Walking and talking the tree of life: Why and how to teach about biodiversity. *PLoS biology, 15*(3), e2001630.

Berat, A. H. İ. (2017). Thinking about digestive system in early childhood: A comparative study about biological knowledge. *Cogent Education, 4*(1), 1278650.

Boersma, K., Waarlo, A. J., & Klaassen, K. (2011). The feasibility of systems thinking in biology education. *Journal of Biological Education, 45*(4), 190–197.

Compton, V., & Compton, A. (2013). Teaching the nature of technology: Determining and supporting student learning of the philosophy of technology. *International Journal of Technology and Design Education, 23*(2), 229–256.

Çuçin, A., Özgür, S., & Güngör Cabbar, B. (2020). Comparison of misconceptions about human digestive system of Turkish, Albanian and Bosnian 12th grade high school students. *World Journal of Education, 10*(3), 148–159.

Dam, M., Ottenhof, K., Van Boxtel, C., & Janssen, F. (2019). Understanding cellular respiration through simulation using lego® as a concrete dynamic model. *Education Sciences, 9*(2), 72.

Dawkins, R. (1996). *The blind watchmaker: Why the evidence of evolution reveals a universe without design.* WW Norton & Company.

Forrester, J. W. (1993). System dynamics as an organizing framework for pre-college education. *System Dynamics Review, 9*(2), 183–194.

Goel, A., Rugaber, S., & Vattam, S. (2009). Structure, behavior & function of complex systems: The SBF modeling language. *International Journal of AI in Engineering Design, Analysis and Manufacturing, 23*(1), 23–35.

Goldstone, R. L., & Sakamoto, Y. (2003). The transfer of abstract principles governing complex adaptive systems. *Cognitive Psychology, 46*, 414–466.

Grant, M. J., & Booth, A. (2009). A typology of reviews: An analysis of 14 review types and associated methodologies. *Health Information & Libraries Journal, 26*(2), 91–108.

Hallström, J., & Klasander, C. (2017). Visible parts, invisible whole: Swedish technology student teachers' conceptions about technological systems. *International Journal of Technology and Design Education, 27*(3), 387–405.

Harsh, M., Bernstein, M. J., Wetmore, J., Cozzens, S., Woodson, T., & Castillo, R. (2017). Preparing engineers for the challenges of community engagement. *European Journal of Engineering Education, 42*(6), 1154–1173.

Hart, E. R., Webb, J. B., & Danylchuk, A. J. (2013). Implementation of aquaponics in education: An assessment of challenges and solutions. *Science Education International, 24*(4), 460–480.

Hmelo, C. E., Holton, D., & Kolodner, J. L. (2000). Designing to learn about complex systems. *Journal of the Learning Sciences, 9*, 247–298.

Hmelo-Silver, C. E., & Pfeffer, M. G. (2004). Comparing expert and novice understanding of a complex system from the perspective of structures, behaviors, and functions. *Cognitive Science, 28*(1), 127–138.

Hmelo-Silver, C. E., Jordan, R., Liu, L., Gray, S., Demeter, M., Rugaber, S., & Goel, A. (2008). Focusing on function: Thinking below the surface of complex natural systems. *Science Scope, 31*(9), 27.

Ho, F. M. (2019). Turning challenges into opportunities for promoting systems thinking through chemistry education. *Journal of Chemical Education, 96*(12), 2764–2776.

Hope, G. (2013). Designing Technology: An exploration of the relationship between technological literacy and design capability. *Design and Technology Education: an International Journal, 18*(2).

Hughes, T. P. (1987). The evolution of large technological systems. *The Social Construction of Technological Systems: New Directions in the Sociology and History of Technology.*

Jacobson, M. J., & Wilensky, U. (2006). Complex systems in education: Scientific and educational importance and implications for the learning sciences. *The Journal of the Learning Sciences, 15*(1), 11–34.

Jung, K., & Otaka, Y. (2019). The introduction of a thin-bending wood speaker as multipurpose teaching material in Japanese junior high school technology classes. *World Journal of Education, 9*(6), 57–64.

Kattmann, U. (2018). A biologist's musing on teaching about entropy and energy: Towards a better understanding of life processes. *School Science Review, 99*(368), 61–68.

Knippels, M. C. P., & Waarlo, A. J. (2018). Development, uptake, and wider applicability of the yo-yo strategy in biology education research: A reappraisal. *Education Sciences, 8*(3), 129.

Magntorn, O., & Helldén, G. (2007). Reading new environments: Students' ability to generalise their understanding between different ecosystems. *International Journal of Science Education, 29*(1), 67–100.

Marton, F. (2006). Sameness and difference in transfer. *The Journal of the Learning Sciences, 15*(4), 499–535.

Marton, F. (2014). *Necessary conditions of learning*. Routledge.

Mella, P. (2012). *Systems thinking: intelligence in action* (Vol. 2). Springer Science & Business Media.

van Mil, M. H., Boerwinkel, D. J., & Waarlo, A. J. (2013). Modelling molecular mechanisms: a framework of scientific reasoning to construct molecular-level explanations for cellular behaviour. *Science & Education, 22*(1), 93-118

Mitcham, C. (1994). *Thinking through technology: The path between engineering and philosophy.* University of Chicago Press.

Ozgur, S. (2013). The persistence of misconceptions about the human blood circulatory system among students in different grade levels. *International Journal of Environmental and Science Education, 8*(2), 255–268.

Park, W. (2020). Beyond the 'two cultures' in the teaching of disaster: Or how disaster education and science education could benefit each other. *Educational Philosophy and Theory, 52*(13), 1434–1448.

Plate, R. (2010). Assessing individuals' understanding of nonlinear causal structures in complex systems. *System Dynamics Review, 26*(1), 19–33.

Rosenkränzer, F., Kramer, T., Hörsch, C., Schuler, S., & Rieß, W. (2016). Promoting student teachers' content related knowledge in teaching systems thinking: measuring effects of an intervention through evaluating a videotaped lesson. *Higher Education Studies, 6*(4), 156–169.

Schooner, P., Klasander, C., & Hallström, J. (2018). Swedish technology teachers' views on assessing student understandings of technological systems. *International Journal of Technology and Design Education, 28*(1), 169–188.

Schooner, P., Nordlöf, C., Klasander, C., & Hallström, J. (2017). Design, system, value: the role of problem-solving and critical thinking capabilities in technology education, as perceived by teachers. *Design and Technology Education, 22*(3), n3.

Senge, P. (2006). The Fifth Discipline: The Art and Practice of the learning Organization. (last edition, revised and enlarged). NewYork: Doubleday/Currency.

Snyder, H. (2019). Literature review as a research methodology: An overview and guidelines. *Journal of Business Research, 104*, 333–339.

Svensson, M., & Ingerman, Å. (2010). Discerning technological systems related to everyday objects: Mapping the variation in pupils' experience. *International Journal of Technology and Design Education, 20*(3), 255–275.

Sweeney, L. B. (2005). How is this similar to that? The skill of recognizing parallel dynamic structures on centre stage. *Creative Learning Exchange: Documents.*

Tripto, J., Assaraf, O. B. Z., Snapir, Z., & Amit, M. (2017). How is the body's systemic nature manifested amongst high school biology students? *Instructional Science, 45*(1), 73–98.

Maria Svensson is a researcher and senior lecturer in the subject area of Technology, University of Gothenburg, Sweden. She is also employed as senior guest researcher at Linnaeus University, Växjö, Sweden. Her research focus is in the field of technology education research with a special interest in technological systems and systems thinking. She also has an interest in technology teacher education and development of science and technology pedagogical knowledge. Two of her latest publications are: Svensson, M., Williams, P., von Otter, A. M., Larsson, J., & Sagar, H. (2021). Technology Content and Concepts in Preschool Teaching–A Practice-based Collaboration. *Techne serien-Forskning i slöjdpedagogik och slöjdvetenskap, 28*(2), 149–155 and Svensson, M. (2021). Systems in Everyday Lives: Making the Invisible Visible. In *Design-Based Concept Learning in Science and Technology Education* (pp. 192–203). Brill Sense.

Chapter 5
Fostering Systems Thinking in the Context of Electronics Studies

Moshe Barak

Abstract Systems thinking is considered a major higher-order thinking skill essential for successful integration in areas such as science, technology, engineering, or management sciences. However, this term has remained somewhat vague and is defined or represented through diverse models or characteristics lists. In the present chapter, we address this challenge by identifying a cluster of systems thinking characteristics related to learning electronics in school and demonstrate how these characteristics correspond with home appliances familiar to any learner: room heating devices, an air conditioner, and a sound system. For example, a heat projector works without feedback (open-loop control); heat diffusers and air conditioners work with negative feedback, often by a thermostat; an unwanted sound may appear in a sound system due to the positive feedback from a microphone placed in front of the speaker. The final section of the chapter underlines four key aspects of systems thinking: modelling, STEM view; the role of the engineer and the technologist; and innovation. Two instructional approaches for fostering systems thinking in the science and technology, constructivist pedagogy and digital pedagogy are also discussed.

Keywords Air conditioner · Feedback control · Modelling · Sound system · STEM view · Systems thinking characteristics

5.1 Introduction

Four major revolutions in the history of technology development can be identified in the modern era (Pouspourika, 2019). The First Industrial Revolution (from 1760) included the invention of the steam engine, and the use of water and steam power to mechanise production. This caused industry to replace agriculture as the backbone of the societal economy. The Second Industrial Revolution (from 1870) included the emergence of new energy sources—electricity, gas, and oil; the creation of the internal combustion engine; the increasing steel demand and chemical synthesis;

M. Barak (✉)
Ben-Gurion University of the Negev, Beer-Sheva, Israel
e-mail: mbarak@bgu.ac.il

J. Hallström and P. J. Williams (eds.), *Teaching and Learning about Technological Systems*, Contemporary Issues in Technology Education,
https://doi.org/10.1007/978-981-16-7719-9_5

and the development of electrical energy and communication methods such as the telegraph and the telephone. Finally, the inventions of the automobile and the plane at the beginning of the twentieth century, which are the reasons why, to this day, the Second Industrial Revolution is considered the most important one. The Third Industrial Revolution in the second half of the twentieth century brought forth the rise of electronics, telecommunications, computers, and the Internet. These new technologies opened the doors to many new fields such as space expeditions, biotechnology, cellular communications, robotics, and high-level automation. According to Schwab (2016), the current Fourth Industrial Revolution (IR 4) is fundamentally changing the way we live, work, and relate to one another. This revolution is a fusion of advanced areas such as artificial intelligence (AI), robotics, the Internet of things (IoT), 3D printing, genetic engineering, and quantum computing.

One can see that electricity and electronics are unquestionably at the heart of the last three technological revolutions mentioned above. Teaching electronics plays an essential role as part of technology education which prepares students for the changing technological world, develop learners' systems thinking and foster their often called twenty-first-century skills such as problem-solving and creativity, critical thinking, collaboration, metacognition, and motivation (Hilton, 2010). However, the traditional way of teaching electronics is no longer relevant (Barak, 2018). In the past, analogue electronics courses started with learning about specific electronics components such as a diode, a transistor, and an operational amplifier; digital electronics courses started with learning about discrete components such as OR, AND, and XOR gates or a digital counter. Today, engineers and students are building electronic and control systems, for example, robotics, using ready-made modules such as power amplifiers, electronic sensors, motors, and digital micro-controllers.

The transition from designing electronic circuits having discrete components to the construction of sophisticated control systems with ready-made modules such as microcontrollers, which occurred developmentally in the market and in education, has created a rational basis for fostering students' systems thinking, for example, understanding aspects of energy, power, feedback loop, and dynamic response of technological systems (Barak & Williams, 2007). However, little work has been done to date to identify a cluster of specific systems thinking principles relevant to high school technology and electronics engineering studies. For example, Yakymenko et al. (2020) refer to STEM as a context for developing systems thinking among electronics engineering students; Chen (2003) addresses systems thinking in the context of using the programmable logic controller (PLC) system; Zuckerman and Resnick (2003) write about a physical interface for system dynamics simulation. Holstien and Bode (2015) describe systems engineering as a technique for applying knowledge from other branches of engineering and disciplines of science in effective combination to solve multifaceted engineering problem. However, all these examples of promoting systems thinking do not classify a structured set of systems thinking aspects relevant to the high school students' world.

In the present chapter, we address this challenge by identifying a set of systems thinking characteristics related to learning electronics in school and demonstrating how these principles correspond with home appliances familiar to any learner: room heating devices, an air conditioner, and a sound system. The final section of this chapter underlines pedagogical aspects of fostering systems thinking in the science and technology class.

5.2 Characterising Systems Thinking

Systems thinking is considered a major higher-order thinking skill essential for successful integration in areas such as science (Orgill et al., 2019; Riess & Mischo, 2010), technology (Hallström & Klasander, 2020), engineering (Frank, 2002), and management sciences (Daellenbach et al., 2012). Although this concept has been discussed for decades in the literature of these fields, it has remained somewhat vague and is defined or represented through diverse models or characteristics lists. In this section, we identify several characteristics of systems thinking that will be examined in the context of teaching electronics.

Chan (2015) writes that a system could mean a set of physical parts of a bigger whole, for example, the structural system of a building or the traction control system of an automobile. Therefore, one aspect of systems thinking involves identifying the parts of a whole or the factors that are important for an outcome. According to Amissah et al. (2020), systems thinking aims at understanding relationships between components and their overall impact on a system's intended and unintended outcomes. Systems thinking is also associated with seeing the 'big picture' or macro-view of things as opposed to a more detailed microscopic view, and it allows one to comprehend how all the pieces fit together to explain how all the parts act to produce the intended effect. Chan (2015) stresses that systems thinking has to do with the ability to 'see the forest from the trees' and solve a problem in a balanced and holistic way, rather than focusing narrowly on only one aspect of the problem. Anderson and Schönborn (2008) associate systems thinking with experts' thinking. Barak (2018) suggested that systems thinking involves identifying and understanding concepts such as the parts and structure of a system; the factors that are important for an outcome; the 'big picture' or macro-view; system boundaries; function and behaviour; feedback in a system; system dynamics; and non-functional properties such as safety and reliability that arise from the interaction between parts of a system.

Hallström and Klasander (2020) suggested the following 11 'clustered groups of concepts' that characterise technological systems: (1) the technical core of a system; (2) hierarchies, sub-systems, components; (3) connections and wholeness; (4) system boundary and surroundings; (5) isolated, closed or open systems; (6) control, feedback and flow of information; (7) system functions and behaviour, processes, models; (8) scale and complexity; (9) dynamics, development, change; (10) socio-technological perspectives; and (11) systems for innovation, conditions for productions.

Arnold and Wade (2015) reviewed seven definitions of systems thinking, including those of Richmond (1994), Senge (1990), Sweeney and Sterman (2000), Hopper and Stave (2008), Kopainsky et al. (2011), Squires et al. (2011), and Forrester (1994).

Arnold and Wade (2015) identified 10 aspects of systems thinking recurring in the various definitions suggested by the above-mentioned researchers. They also counted how many researchers out of seven included each aspect in their model, as displayed in parentheses in the following list: Whole rather than parts (4); Interconnections/interrelationships (4); Feedback loops (4); Dynamic behaviour (4); System as the cause of its behaviour; System structure generates behaviour (2); Stock and flow relationships (2); Nonlinear relationships (2); Delays (1); and Acknowledgment that systems are important (1).

In summary, we have seen several lists of aspect of technological systems and systems thinking, even though they overlap to some extent. In the following sections of this chapter, we will address the following seven key characteristics of systems thinking derived from a synthesis of the just cited literature and most relevant to electronics teaching:

W—Whole rather than the parts: recognising system properties that are the outcome of the interaction between system ingredients and cannot be related to a specific component.
I—Interconnections/interrelationships: understudying the flow of information and energy in a system.
F—Feedback loops: realising the process of setting a desired value to a controlled variable, measuring the actual value in real time, and regulating system operation to reduce deviation.
D—Dynamic behaviour: observing change of a variable in a system, for example, an increase or decrease, which depends on time.
B—System boundaries: identifying the border between internal processes and external entities that may affect the system but are unaffected by it.
T—Technological innovation: realising the invention of new products or services that answer needs or may push the market forward.
S—Socio-technological implications: appraising social, economic, or environmental implications of a system.

Figure 5.1 shows that the seven characteristics of systems thinking mentioned above partially overlap and that there is no built-in hierarchy between them. For example, seeing the 'whole rather than the parts' is associated with identifying 'interconnections and interrelationships' and understanding 'socio-technological implications' of a system. It is also worth noting that not all of these principal aspects are necessary to describe a system, but they depend on the complexity of the system in question.

In the following sections, we will explain and demonstrate how the seven characteristics of systems thinking W, I, F, D, B, T, and S presented in Fig. 5.1 are expressed in three technological systems—room heating devices, an air conditioner, and a sound system, and how to teach this subject in order to foster systems thinking in these contexts.

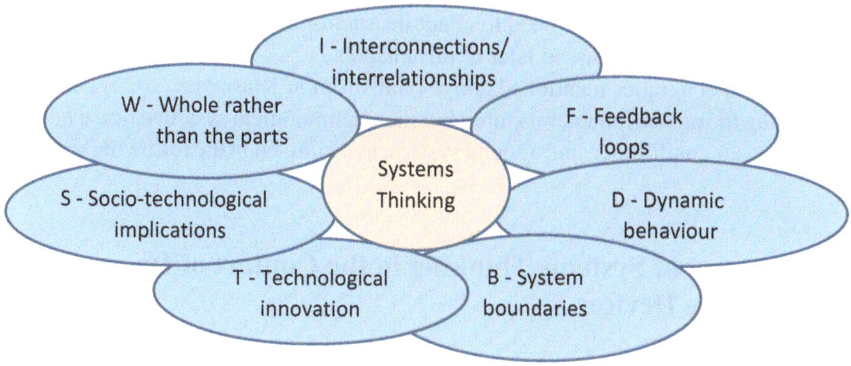

Fig. 5.1 Principal characteristics of systems thinking in a technological context

5.3 Pedagogy for Teaching Aspects of Systems Thinking

It is widely agreed that the traditional teacher-centred instruction contributes very little to fostering higher-order cognitive skills such as systems thinking. Consequently, education today aims at implementing student-centred instruction in class derived from the constructivism learning theory. Constructivist concepts of learning, which have their historical roots in the work of Dewey (1929), Bruner (1961), Vygotsky (1962), and Piaget (1980), call to engage students in problem-solving and critical thinking regarding authentic learning activities that learners find relevant and engaging (Briner, 1999; Brown et al., 1989). Therefore, it is desirable to teach systemic thinking in the context of technological systems with which students are familiar with everyday life. Hallström and Klasander (2020) suggest more specific pedagogies for teaching technological systems to secondary school students:

1. **Interface pedagogy**: starting with the interface between the supposed system and the human being using it, for example, exploring the hidden system in a smartphone or ATM machine.
2. **Holistic pedagogy**: starting with a known technological system, for example, the railway, and moving from the wholeness to identifying sub-systems and significant components.
3. **Historical pedagogy**: following the historical development or changes in a system, for example, the telephone, from a previous point in time to the present, or vice versa.
4. **Design pedagogy**: designing, 'making', and improving a product or a technological system of appropriate complexity (pp. 73–74).

The instructional pedagogies suggested by Hallström and Klasander go hand-in-hand with the broad notion of constructivist pedagogy that has played a key role in the educational literature of recent decades. However, educators should be aware of the limitations and difficulties in implementing students-centred instructional methods in school, and especially the need to adapt instruction to students' prior knowledge and skills, as will be discussed later in this chapter.

The four pedagogies mentioned above (Hallström & Klasander, 2020) guided us in focusing the present chapter on three familiar technological systems: home heating devices, an air conditioner, and a sound system, as discussed in the following sections.

5.4 Aspects of Systems Thinking in the Context of Room Heating Devices

The first example for promoting systems thinking in electricity and electronics we present in this chapter relates to two different room heating devices, the heat projector shown in Fig. 5.2 and the hot air diffuser shown in Fig. 5.3.

Let us look at these devices considering the seven aspects of systems thinking we are addressing in this chapter (Fig. 5.1).

5.4.1 Seeing the 'whole Rather Than the Parts' of Heating Devices

Seeing the 'whole rather than the parts' relates to understanding the purpose of a system, its main components, how it works, and its suitability to user needs. This aspect is central to systems thinking, as has been mentioned by many researchers (Arnold & Wade, 2015; Chan, 2015; Frank, 2002). Comparing two or more similar products (as we often do in a store) and listing their pros and cons may help in understanding the features of each system.

Fig. 5.2 Heat projector

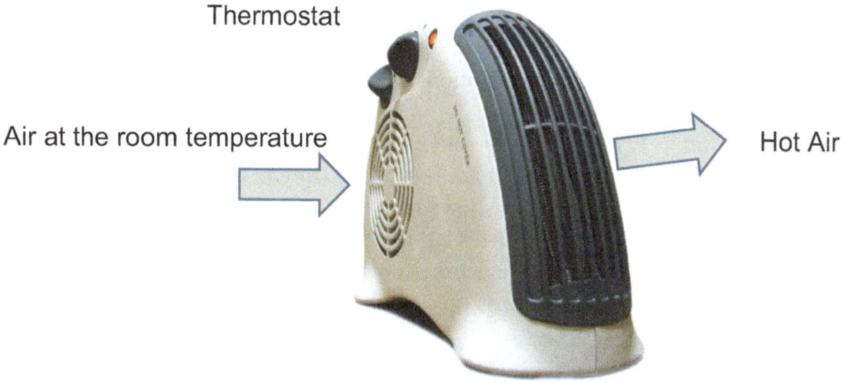

Fig. 5.3 Hot air diffuser with a thermostat

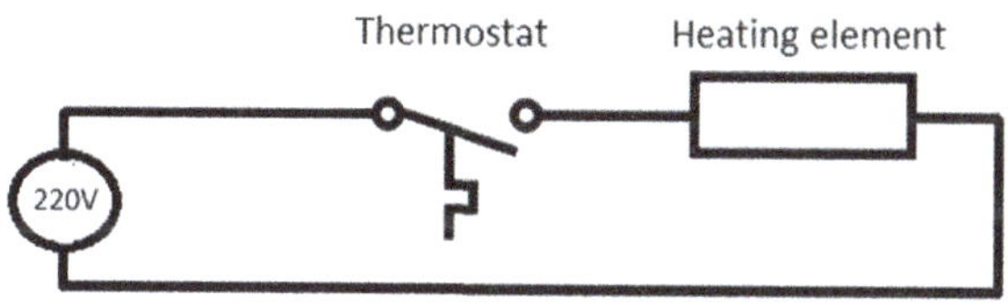

Fig. 5.4 Thermostat control over a heating element

The heat projector main components are heating two elements and on/off switches. The device converts electrical energy into heat energy.

Pros: Quickly heats the air around the device.

Cons: Keeps heating even when the temperature is very high. Could cause a fire.

The main components of the hot air diffuser are a heating element, a blower that draws air at room temperature and dissipates it back to the room at a higher temperature, and a thermostat that stops heating when the temperature reaches the desired value, as shown in Fig. 5.4. The device also converts electrical energy into heat energy.

Pros: Disperses heat over a wide area; stops heating when the air temperature reaches the desired value.

Cons: Does not provide powerful heating near the device.

5.4.2 Interconnections/Interrelationships Between Components in Heating Devices

To identify interconnections or interrelationships in a system, one needs to understand how the main components in the system relate to each other, and how information signals and energy flow in the system. In the case of the heat projector, we can describe the flow of energy as follows:

Electrical supply (110 or 220 V) > manual power switch > heating element > heat radiation to the environment.

The interconnections and interrelationships related to the hot air diffuser could be described as follows:

Electrical supply (110 or 220 V) > automatic thermostat switch > heating element > heating air streamed created by a fan > heat diffusing to the environment.

The textual presentation above shows the structure and interconnections of the components in the two heating devices but indicates only indirectly the most important difference between the two electrical devices—the fact that only the hot air diffuser included a feedback loop, as presented in the following section.

5.4.3 Feedback Loop with a Thermostat

The thermostat is the heart of the hot air diffuser control system because it:

- is used to set up the desired temperature by the user,
- measures the actual temperature of air in the room,
- turns the heating element on/off according to the measured temperature.

In the professional language, this is called a feedback loop or a feedback control system, and it is usually described by a block diagram, as shown in Fig. 5.5.

In Fig. 5.5, the minus sign at the feedback signal indicates that when the temperature in the room increases, the thermostat turns the heating off, and vice versa. In other words, the negative feedback stabilises the controlled variable—the temperature.

Although many home or industrial heating or cooling appliances use a mechanical thermostat, this device is frequently replaced by modern electronic sensors and control systems. However, all of them follow the same control process described above.

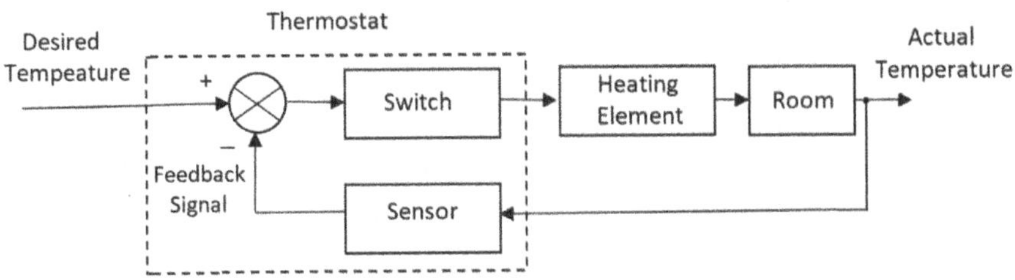

Fig. 5.5 . Block diagram of a thermostat's temperature control system

5.4.4 Dynamic Behaviour of Temperature in a System

The term dynamic behaviour relates to a description of how a system or an individual unit functions with respect to time (McGraw Hill Dictionary of Scientific & Technical Terms, 2003a, 2003b). Barak and Williams (2007) noted that at the heart of systems thinking in scientific, technological, and social contexts lies the analysis of the change with time of physical or social variables, such as the temperature of air or a mass, the volume of water in a tank or the number of products in a warehouse. Analysis of dynamic processes frequently relates to the instantaneous value of a specific variable, its rate of change (derivative), or accumulation (integral) with time. Hallström and Klasander (2020) note that system dynamics is one of the most important characteristics of technological systems.

In referring to temperature control by a thermostat, we have seen that when the air temperature in the room reaches the desired value, the thermostat turns the current to the heating element off; when the temperature drops below the desired value, the thermostat turns the heating on. To avoid turning the heating on and off too frequently, the thermostat has a 'cyclical' range (also called a hysteresis range) of several degrees. For example, as shown in Fig. 5.6, if the required temperature is set to 20 °C (set point) the thermostat holds the heating on until the actual temperature reaches $T1 = 21$ °C and turns the heating off until the temperature decreases to $T2 = 19$ °C.

Regarding the process described in Fig. 5.6, the larger the hysteresis range $T1$-$T2$, the longer the thermostat switching times are on/off. It should be noted that a rise or fall in temperature does not occur at a constant rate but according to an exponential function, according to which the rate of change (rise or fall) decreases depending on time (Barak & Williams, 2007). In summary, we have seen that in temperature controlled by a thermostat, we do not get a constant temperature at the desired value but a dynamic response in which the temperature rises and falls cyclically around

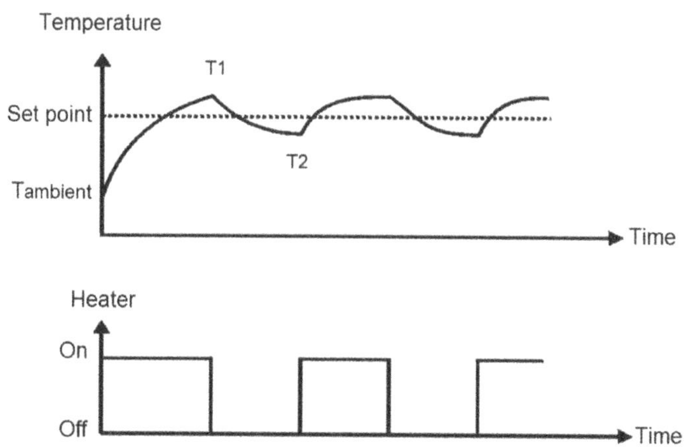

Fig. 5.6 Temperature control by a thermostat with a cycling range

the desired value. This phenomenon also exists in other temperature control systems using a thermostat, such as a baking oven, refrigerator, or air conditioner, as shown in the next section.

5.4.5 System Boundaries

We saw that the term system boundary refers to the border or distinction between internal and external objects or processes in a system. According to Lavi and Dori (2019), external entities of a system are objects and processes that affect the system but are unaffected by it. Internal objects and processes affect system operation and are affected by external entities. With respect to a heat projector and hot air diffuser, external components are, for example, the structure of the room, the outside temperature, and the thermal isolation of the windows, walls, and ceiling. These entities could affect the temperature in a room but are not affected by the heating or cooling devices.

5.4.6 Innovation

The heat projector and hot air diffuser are rather simple devices and have relatively little room for innovation. An advanced heat diffuser may include, for example, a mechanism that rotates the appliance to the right/left for optimal hot air distribution, a water tank to increase humidity in the room, a timer for automatic shut-off after a limited time, and a safety switch to stop heating in case the appliance overturns.

5.4.7 Socio-Technological Implications

The heat projector and hot air diffuser both convert electrical energy into heat energy and are considered heavy electricity consumers. Not only does the user pay a higher price for the electricity consumption of these devices, the devices also contribute to environmental pollution due to electricity production in power stations.

5.4.8 Summary

Up to this point, we have used the examples of a heat projector and a hot air diffuser to illustrate four of the seven concepts related to systems thinking that we are focusing on in this chapter (Fig. 5.1): seeing the whole rather than the parts; interconnection/interrelationships between system components; a feedback loop by the thermostat; and dynamic behaviour of the controlled variable. In the following sections, we will address further aspects of systems thinking in the context of technological systems and discuss how to teach systems thinking in class.

5.5 Aspects of Systems Thinking in the Context of an Air Conditioner

The second example of fostering systems thinking in the context of learning electronics engineering is the case of the air conditioner, a complex electro-mechanical device found today in many homes and offices. We chose to present this example because it combines aspects of energy conversion, control, electricity, electronics, mechanics, and computing. Below we will examine the air conditioner system from the perspective of the seven aspects of systems thinking we are focusing on in this chapter (Fig. 5.1).

5.5.1 Seeing the Whole Rather Than the Parts

A modern air conditioner can be used for cooling or heating a room or larger spaces. Our explanation relates to a split-system air conditioner, as shown in Fig. 5.7.

Regarding the air conditioner, seeing the whole rather than the parts means understanding that when the internal part of the air conditioner draws air from the room and cools it, the heat energy is transferred to an external device, which dissipates it outside. The air conditioner does not transfer air from the room outside, or vice versa.

Fig. 5.7 Split-system air conditioner

5.5.2 *Interconnections/Interrelationships in an Air Conditioner*

To understand the interconnections/interrelationships between components in an air conditioner, one must identify the main components of the system and how they are connected or related to each other, as shown in Fig. 5.8.

Figure 5.8 shows that the main components of the air conditioning system are:

- Internal radiator,
- External radiator,
- Compressor controlled by a thermostat,
- Refrigerant gas,
- Expansion valve.

Understanding the function of each component and their interconnections or interrelationships is a key in understanding how the air conditioner system works.

From physics theory, we know that when a liquid is converted into a gas in a process called phase conversion, it absorbs heat. When a gas is converted into a liquid by high pressure, it emits heat. Air conditioners exploit this feature of phase conversion by forcing a special gas to evaporate and condense periodically in a closed system of coils. The compressor drives the gas in a circular manner between the two radiators:

- On the evaporator side (internal radiator), the compressor sucks in the gas, i.e. creates a low pressure. The gas draws heat from the environment, i.e. emits cold air.
- On the condensing side (external radiator), the gas is compressed to a high pressure and is converted into a liquid. In this process, the radiator emits heat to the environment.

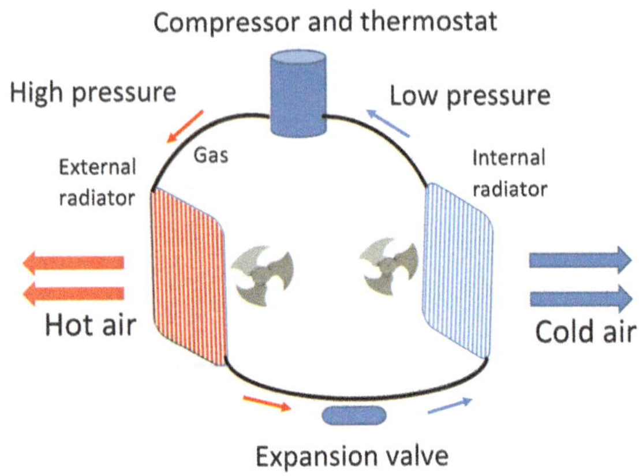

Fig. 5.8 Main components of an air conditioning system

The expansion valve plays an important role in the process described above because it causes the gas to change from a high pressure to a low pressure, as shown in Fig. 5.8.

It is worth mentioning that modern air conditioners are also designed to heat a house on cold days. For this purpose, the gas system in the air conditioner includes a 'reverse valve' (not shown in Fig. 5.8) that reverses the gas flow direction between the two radiators so that the inside radiator is the condenser, which emits heat to the room, and the outside radiator is the evaporator unit, which absorbs heat from the environment.

In summary, understanding the interconnections/interrelationships in an air conditioner requires knowing keywords related to an air conditioning system, such as internal and external radiators, gas, compressor, and thermostat. More challenging is understanding the process of gas condensation and evaporation that causes the emission and absorption of heat energy.

The compressor, which does the 'hard work' of an air conditioner, includes an electrical motor and a mechanical compressor, as shown in Fig. 5.9.

In Fig. 5.9, we can see on the right side of the compressor the connection points of the pipes for the gas input and output to/from the compressor. On the left side are the wire connections for the electricity supply. The electric power supplied to the compressor motor determines the intensity of the air conditioner. For example, an air conditioner of 1 HP (736 W) power can cool or heat a relatively small room inside an apartment.

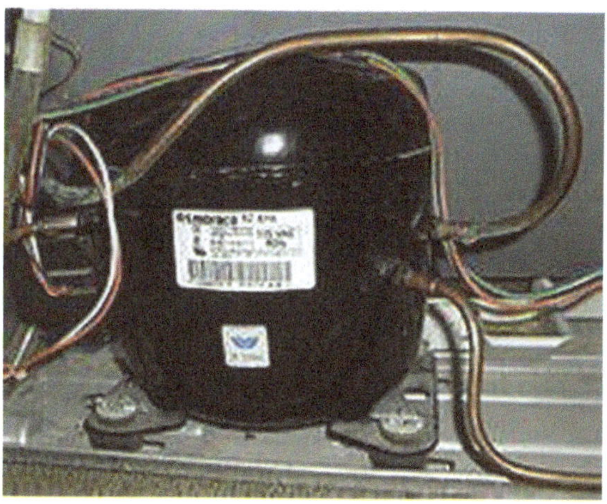

Fig. 5.9 Air conditioner compressor

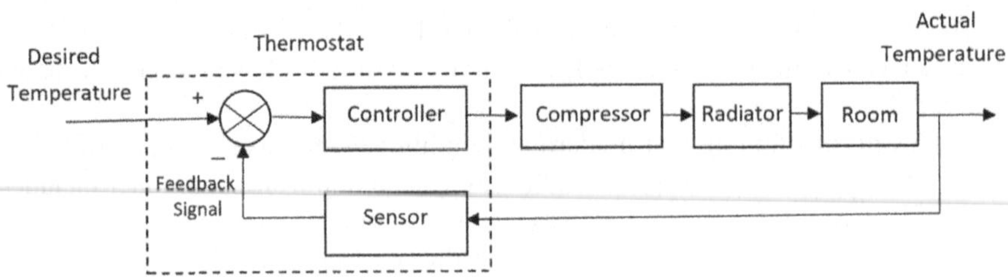

Fig. 5.10 Block diagram of an air conditioner control system with a thermostat

5.5.3 *Feedback Loop with a Thermostat*

Most air conditioners today are controlled by either a mechanical or electronic thermostat. Like in the example of a hot air diffuser we saw in the previous section, the air conditioner thermostat does the three basic actions of a feedback system:

- **Measures** the actual value of the controlled variable—the temperatures.
- **Compares** the measured values to the desired value.
- **Amends/regulates** the process to achieve the desired value.

As we saw in the previous example of temperature control by a heat diffuser, it is common in technology and engineering to represent a system with a block diagram that shows the main components and signals in the system, as shown in Fig. 5.10.

The minus (−) sign near the feedback signal arrow indicates 'negative feedback': when the measured temperature increases too much, the thermostat turns the compressor on to cool the room; when the temperature decreases too much, the thermostat turns the compressor off to stop cooling the room. In other words, the negative feedback stabilises the temperature in a required range.

5.5.4 *Dynamic Behaviour*

In the previous example relating to the hot air diffuser, we have seen that for temperature control by a thermostat, the temperature in the room is not constant but rises and falls in cycles (Fig. 5.6). A similar phenomenon also exists in air conditioner operation. For example,

- A customer adjusts the thermostat to a desired temperature of 24 °C (set point).
- If the temperature in the room is too high, the air conditioner cools the room until the temperature drops slightly below the desired value, for example, 23 °C, and stops cooling.
- The temperature in the room gradually rises until it reaches a level slightly above the desired value, for example, 25 °C, then the air conditioner starts running again and cools the room.

In this case, a graph illustrating the change in temperature versus time will resemble Fig. 5.6, but for the opposite situation: when the air conditioner is on, the temperature drops, and when the air conditioner is off, the temperature rises.

Most air conditioner users are unaware of the phenomenon of cycling described above, also called 'hysteresis' range. However, this phenomenon causes the air conditioner compressor to work for a few minutes and disconnect periodically, causing energy wastage and wear-and-tear of the air conditioner motor. Below we will present an innovative method for controlling an air conditioner that prevents this phenomenon.

5.5.5 System Boundaries

We have seen that the term system boundary refers to the distinction between internal and external objects or processes in a system. In the case of an air conditioner, like a room heater, external factors include the structure of the room, the outside temperature, and the thermal isolation of the building. These entities might affect the temperature in a room but are not affected by cooling or heating by the air conditioner.

5.5.6 Technological Innovation in Air Conditioning Systems

In recent years, technologists and engineers around the world have been investing efforts in refining conventional air conditioners we have known for decades. Below is an example of these developments.

So far, we have seen that the motor of a conventional air conditioner is turned on/off periodically by a thermostat, and the temperature in the room rises and falls cyclically, as illustrated in the blue lines shown in the two graphs in Fig. 5.11. In recent years, modern air conditioners have been increasingly using the electronic inverter technology to substitute thermostat control. The inverter electronic controller provides the motor with a high current only during start-up and then runs the motor continuously at steady low power and low speed, as illustrated in the red lines in Fig. 5.11. This process replaces the thermostat on/off control method, as marked in the blue lines in Fig. 5.11.

Studying the implications of the transition from thermostat-based control to inverter technology-based control of an air conditioner directly relates to the aspect of dynamic behaviour of a technological system. As explained earlier, in a system based on the traditional technology of the thermostat, the air conditioner's compressor is turned on and off periodically and the temperature rises and falls. In a system based on inverter technology, the motor works continuously at low intensity and the temperature in the room (or refrigerator) is constant, without causing fluctuations in room temperature.

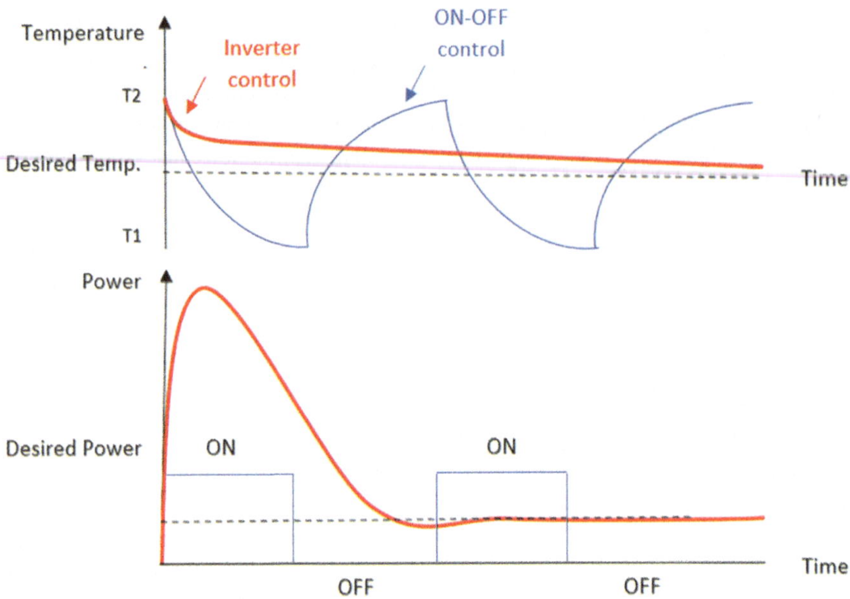

Fig. 5.11 Air conditioner based on electronic inverter control (red lines) versus on/off thermostat control (blue lines) (Izushi, 2018)

The use of inverter technology is expanding to other household appliances such as refrigerators and washing machines. The benefits are not only a smarter and quieter operation of these devices but also low energy consumption. The disadvantages are that inverter-based devices are often more expensive compared to thermostat-based devices, and the electronic circuit may be more fault-sensitive compared to a mechanical thermostat. However, it is likely that these points will improve in the future.

5.5.7 *Socio-Technological Aspects of Room Heating Devices and Air Conditioners*

De Vries (2005) describes technology as 'the human activity that transforms the natural environment to make it fit better with human needs, thereby using various kinds of information and knowledge, various kinds of natural (material, energy) and cultural resources (money, social relationships, etc.)'. This definition emphasises that systems thinking is about understanding a technological system not only from scientific or engineering aspects but also from the aspects of human needs and aspirations.

In the present study, we have seen that using a modern air conditioner based on inverter technology instead of the old on/off control method may reduce fluctuations in air temperature, reduce noise, and save energy. From another perspective, air conditioner producers are now required to use refrigerant R-32 because it can reduce electricity consumption compared to that of air conditioners using refrigerant R-22 or R-410. Moreover, R-32 has a global warming potential (GWP) that is one-third lower and is remarkable for its low environmental impact.

Although air conditioners consume much less electricity than other heating devices, air conditioners are still among the most significant energy consumers at the home or the workplace and thus contribute to increased energy production and environmental pollution. To deal with this challenge, the 'green home' movement (Bonta & Snyder, 2009) recommends designing houses that conserve energy or water, improve indoor air quality, use sustainable, recycled or used materials, use more energy-efficient appliances or building materials that are more efficient in managing temperature. The 'smart home' energy network (Han & Lim, 2010) has gained widespread attention due to its flexible integration into everyday life by effectively delivering solutions for diverse areas including consumer electronic device control, energy management, and efficiency home and commercial building automation, as well as industrial plant management.

Up to now, we have seen that room heating devices and air conditioners may serve as rich examples for teaching all attributes of systems thinking to students. The question is which pedagogy is best suited to achieve this goal, as will be discussed in the following section.

5.5.8 Pedagogy for Fostering Systems Thinking in the Context of Room Heating Devices and Air Conditioners

As noted in the Literature Review section of this chapter, educators strongly recommend using a constructivist pedagogy in teaching systems thinking in contrast to the teacher-centred traditional pedagogy. Student-centred pedagogy could be represented by the following four principles: contextual learning; activity-based learning; collaborative learning; and reflective learning. To achieve this end, teachers can engage students in problem-, project-, or design-based learning (Barak, 2020). Below are examples of tasks for students aimed at developing systems thinking in the context of air-conditioning systems while noting the main systems thinking aspects that are most relevant to each task out of W, I, F, D, B, T, S.

1. Investigating the process of choosing an air conditioner for different spaces (main systems thinking aspects: W, B, T, S).
2. Researching the range of temperature change (maximum/minimum) in a room and the frequency of the thermostat on/off switching by a data recorder (main systems thinking aspects: W, I, F, D, B).

3. Exploring the impact of having an open window in a room on air conditioner performance and energy consumption (main systems thinking aspects: W, I, F, D, B, S).
4. Comparing the effectiveness of air conditioners based on mechanical thermostat versus invertor technology in terms of dynamic response and energy consumption (main systems thinking aspects: W, I, F, D, B, T, S).
5. Designing and constructing a micro-controller-based electronics system to monitor and control temperature and humidity in an air-conditioned space (main systems thinking aspects: W, I, F, D, B).
6. Learning from experienced air conditioner installers about issues such as how to choose, install and maintain this device, product quality, common faults in air conditioners, and how to repair them (main systems thinking aspects: W, I, F, D).

In general, it could be said that each of the tasks listed above corresponds with all seven aspects of systems thinking W, I, F, D, B, T, S. Nevertheless, we chose to mark in each task aspects of systems thinking that might have a slightly higher weight in the same task. An additional viewpoint for fostering systems thinking in the technological and electronic class in the context of air conditioners is the four pedagogies for fostering systems thinking proposed by Hallström and Klasander (2020), as noted in the Literature Review section: interface pedagogy; holistic pedagogy; historical pedagogy; and design pedagogy. These steps might complement and take forward the six examples of tasks for students mentioned above.

In summary, we have seen several major characteristics of systems thinking related to heating and air conditioning devices, how to engage students in learning these systems, and thinking patterns in the technological and science class.

5.6 Aspects of Systems Thinking in the Context of a Sound Amplification and Enhancement System

In this section, we discuss how systems thinking relates to systems for sound amplification and enhancement, with a focus on the seven aspects of systems thinking W, I, F, D, B, T, S we are discussing in this chapter (Fig. 5.1).

5.6.1 Seeing the Whole Rather Than the Parts

A simple example of a sound amplification system that includes a microphone, amplifier, and loudspeaker is shown in Fig. 5.12.

Seeing the whole rather than the parts in this system means recognising the role of the system, its main elements, and how it works in general. From the user's viewpoint, the system includes microphones, some electronic instrumentation inside

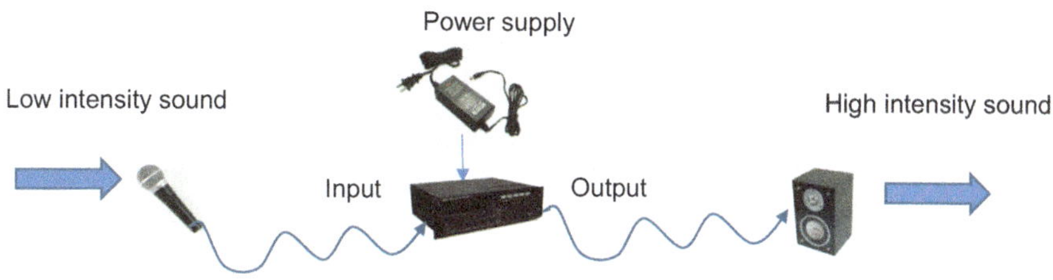

Fig. 5.12 A basic sound amplification system

the box and a loudspeaker. In this case, seeing the whole means understanding that the microphones convert the voice to an electronic signal, which the box amplifies. It also includes an intuitive understanding that the singer's voice spreads through the air, for example, to the microphones and from the speaker to a listener. Many people are familiar with the term sound waves, which is somewhat vague. Figure 5.13 shows a simulation of sound waves created by a loudspeaker.

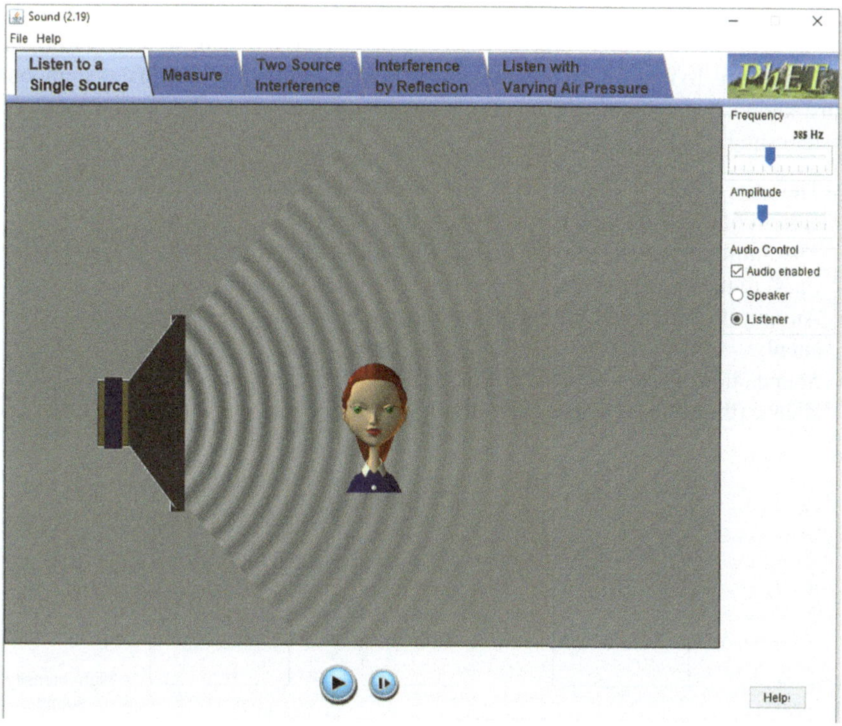

Fig. 5.13 Simulation of sound waves. *Source* PhET free interactive simulations, University of Colorado Boulder https://phet.colorado.edu/. *Licensing:* https://phet.colorado.edu/en/licensing

5.6.2 *Interconnections/Interrelationships in a Sound Amplification System*

This aspect of systems thinking takes us one step further in understanding the phenomena associated with a sound system physically and technologically.

From physics, we know that sound happens because something makes air (or other material) molecules vibrate. When we strum a guitar string, the string oscillations vibrate the air molecules around it, which vibrate other molecules nearby, and so on. Each particle vibrates locally but the vibrations propagate in the air, a phenomenon that we call a sound wave, which is shown in Fig. 5.13.

From the technological perspective, sound systems deal with transforming sound to an electronic signal, amplification of this signal and transforming it back to sound, as is shown in Fig. 5.12.

As we have seen in the air conditioner section, it is common in technology and engineering to represent a system with a block diagram that shows the flow of signals and energy in a system. The sound amplification system could be represented by a three-stage block diagram, as shown in Fig. 5.14.

In the block diagram, each of the three main blocks represents a component (or sub-system) in the system, and between the blocks are the main signals (variables) in the system. To understand the interconnections/interrelationships in a system, one must know a little about the signals characterising the system. For example, in the system shown in Fig. 5.14:

- The microphone might provide an electrical signal of 1 millivolt (mV) and a current of a few microamperes (uA) to the amplifier.
- The amplifier would provide a signal of 1 V and a current of a few milliamps (mA) to the loudspeakers.
- An amplifier requires an energy source such as batteries or an electric power supply.
- Sound intensity is measured in dB (deci-Bell) (Berg, 2020). For example, absolute silence (threshold of hearing)—0 dB; home/office background noise—50 dB; rock

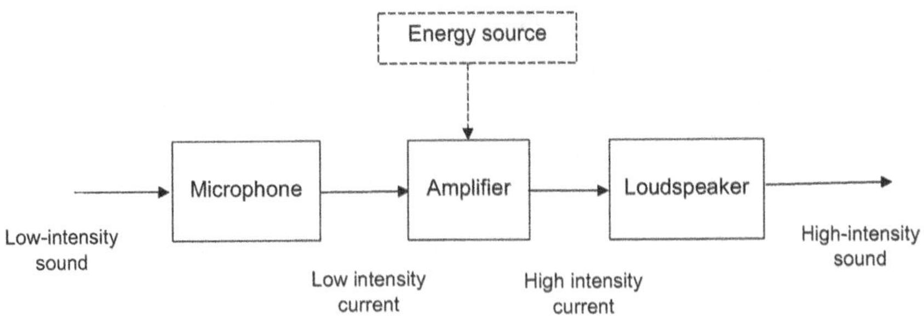

Fig. 5.14 Block diagram of a sound amplification system

Fig. 5.15 A sound intensity metre

music—120 dB. Sound intensity could be measured by a sound metre, as shown in Fig. 5.15.

It is important to note that the term sound system has to do with the amplification and enhancement of sound for many uses such as theatre and musical performances, or a public address system. In these cases, a sound system could be quite complex and include an advanced power amplifier and several loudspeakers of different sizes. Such a system often includes a 'mixing console' that enables connecting to the amplifier sound from different devices such as singers' microphones, electronic guitars, a CD player, or a computer. The sound console is the major tool for the sound person to adjust the intensity and enhance the quality of sound from each input device (Fig. 5.16).

Drawing a block diagram for an advanced sound system as described above is a challenge for the teacher and the student.

5.6.3 Feedback Loops: Unwanted Sounds Due to Positive Microphone Feedback

When we use a sound amplification system such as a karaoke system, at first glance this looks like an open-loop system, that is, without feedback. However, an unwanted high-pitched jarring sound is often obtained when the microphone is placed in front of

Fig. 5.16 Sound console

the speaker, as shown in Fig. 5.17. This phenomenon is called microphone feedback. In this case, the microphone signal is amplified through the amplifier and speaker, and the sound from the speaker is picked up by the microphone and then amplified again.

A block diagram of the microphone feedback system is shown in Fig. 5.18.

The phenomenon of microphone feedback is defined as 'positive feedback', which causes instability or self-oscillations in a system. In contrast, we have seen earlier in this chapter the case of 'negative feedback', which stabilises the temperature in a room controlled by a hot air diffuser or air conditioner (Figs. 5.5 and 5.10).

Positive feedback from the microphone to the speaker occurs when a sound signal repeats itself and is received in the microphone in the same phase each cycle. From a physical viewpoint, this happens only if the microphone is placed at exactly one

Fig. 5.17 Microphone feedback that causes unwanted sounds

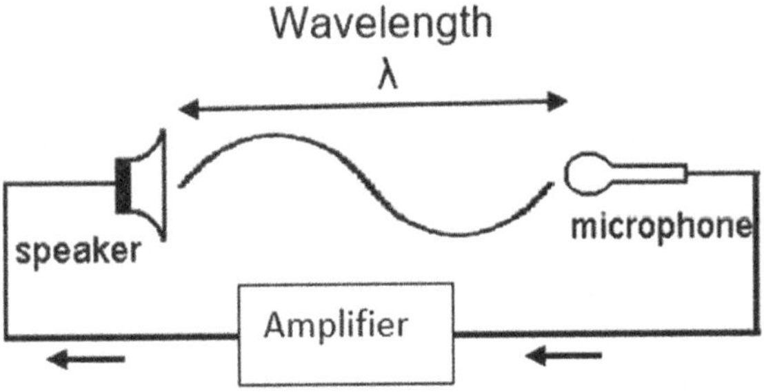

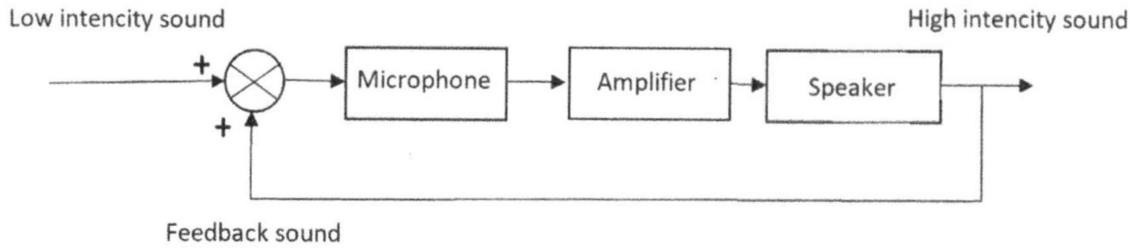

Fig. 5.18 Block diagram of unwanted positive feedback from a microphone

wavelength λ (or a whole number of wavelengths n•λ) from the loudspeaker, as shown in Fig. 5.17.

From physics, we know that sound wavelength is calculated by the formula $\lambda = V/f$ (m), where V (m/s) is the sound velocity of propagation and f (Hz) is the sound frequency. For example, the sound propagation in air is $V = 343$ m/sec, and for a sound of 1 kHz, the wavelength is $\lambda = 34.3$ cm.

If the microphone is placed at distance of 34.3 cm from the amplifier, an unwanted sound will appear at a frequency of 1 kHz. If we move the microphone closer to the amplifier, the sound frequency will increase. A simple way to avoid microphone feedback is to change the location of the microphone. Professional sound people have practical experience in how to resolve this issue.

The phenomenon of positive feedback that causes instability or unwanted oscillations is a major challenge not only in a sound system but also in designing control systems such as robots or airplanes. Regarding the characteristics of systems thinking, this issue clearly belongs to the aspect of understanding feedback loops.

5.6.4 *Dynamic Behaviour: Frequency Response of a Sound System*

Dynamic behaviour of a sound system is associated with its frequency response, namely the audible frequency range that a sound system can reproduce. Audio frequencies are measured in Hertz (Hz), and the theoretical range of human hearing is from about 20 Hz (the lowest bass tones) to 20 kHz (the highest treble tones). Frequencies at the low end of this range are typically referred to as bass, the higher end is called treble, and those in the middle are called mid-range. However, considerable variation exists between an individual's hearing, especially at high frequencies, and a gradual loss of sensitivity to higher frequencies with age is considered normal.

Good speakers are the key to getting high-quality audio from a sound system. Ideally, a single speaker would be able to play sounds at all frequencies 'equally', meaning it would not make certain frequencies louder than others would. Typically, a sound system and the speakers create lower intensity sounds at low frequencies, for example, below 100 Hz, and at higher frequencies, for example, above 15 kHz, as shown in Fig. 5.19.

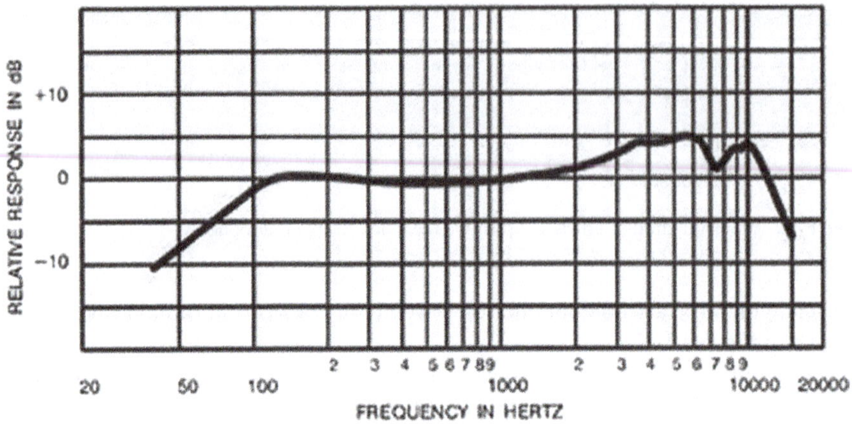

Fig. 5.19 An example of frequency response of speakers

In the case of a sound system, the aspect of system dynamic behaviour is crucial. In practice, it is very important to check the frequency range of a sound system or headphones that we are considering purchasing.

5.6.5 System Boundaries: Noise Accompanying a Sound System

As previously mentioned, system boundaries are defined by external objects and processes that affect the system but are unaffected by it (Lavi & Dori, 2019). In a sound system, the main factor that defines system boundaries is the noise from external resources that accompanies the sound, as shown in Fig. 5.20.

The phenomenon of unwanted noise is a major problem in electronics systems, for example, in radio and television broadcasts or digital communication networks. Common external noise sources are low-quality wires or cords, loose contacts, and noise from the 50/60 Hz power system that adds to the sounds. One can find websites on the Internet that advise how to eliminate noises in sound systems, it is often impossible to eliminate noise completely.

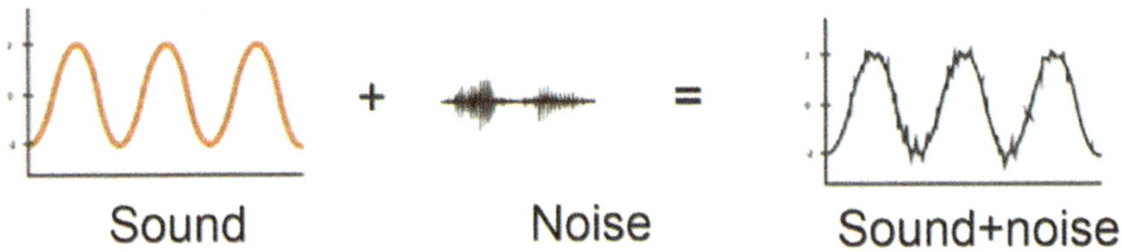

Fig. 5.20 Sound with noise

5.6.6 Technological Innovations in Sound Systems: Active External Sound Cancellation

Innovation is at the heart of technological development. The sound system we purchase today is much more sophisticated than what we knew 10 or 20 years ago. Over the years, new components and features have been added to sound systems, such as using a remote control, connecting the sound system to a TV using an optical audio cable, or listening to music on our smartphone by Bluetooth connection to headphones. All these enhancements have been developed by the initiative, creativity, and imagination of engineers and technologists. In many cases, consumers get to these innovations quickly and start thinking about how they had managed without them before. One of the innovations in sound systems is an active noise-cancelling technology, as shown in Fig. 5.21. This system includes a microphone that detects incoming noise and introduces a second opposite sounds in sync with the problem noise. This has the effect of reducing the problem of external noise.

Another area of innovation in the field of sound is that of hearing aids. A traditional hearing aid is based on analogue electronic technology and includes a microphone, amplifier, and tiny speaker, like the audio system we have seen. In recent years, hearing aids based on digital technology have been developed. The tiny digital processor installed in the hearing aid makes it possible to better adjust the sound amplification to frequencies where there is a hearing impairment, increase speech

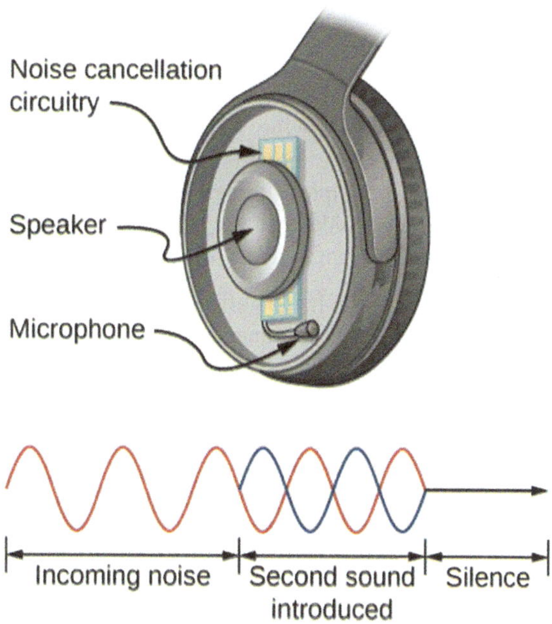

Fig. 5.21 Headphones with active noise cancellation

sounds, mute background noise, and eliminate beeps due to positive feedback from the microphone.

5.6.7 Socio-Technological Aspects of Sound Systems

Sound has played a key role throughout the history of humankind, including hearing sounds from nature, animals, human speech, and music. The technology of recording and re-playing sound has evolved since the invention of the vinyl disk in 1897. The use of magnetic tape for sound recording originated in the 1930s, and the compact disc (CD), which is a digital optical technology for data storage, was invented in 1982. Today, we are experiencing the revolution of storing huge amounts of data, including sound and movies, on digital databases, and listening to sound or watching movies online anywhere, anytime. It is a cultural revolution that affects not only the listeners but also the artists who create music and movies of all kinds. It would be interesting to explore aspects of systems thinking related to new technologies for sound storage and playing but lying at the end point of the new systems is still the fundamental sound amplification system discussed in this chapter.

5.6.8 Pedagogy for Fostering Systems Thinking in the Context of Sound Systems

As previously noted, educators strongly recommend moving from teacher-centred traditional pedagogy to student-centred constructivist pedagogy, which could be described by the following four principles: contextual learning; activity-based learning; collaborative learning; and reflective learning. Below are examples of applying these principles through problem-, project-, or design-based learning to cultivate systems thinking in the context of sound systems. For each example, we mark the most relevant systems thinking aspects of W, I, F, D, B, T, S.

1. Investigating the factors affecting the quality of sound in a stage performance, for example, the components of the sound system; the role of the sound mixing console; the process of adapting a sound system to the environment of a stage performance or to individual singers, actors, or a musical band; observing the work of a professional sound person (main systems thinking aspects: W, I, B).
2. Exploring the quality of sound produced by a common home sound system; measuring the sound intensity depending on the frequency; drawing a graph of the system frequency response; identifying noise sources in the system (main systems thinking aspects: W, I, D, B).
3. Designing and constructing a digital micro-controller-based device that measures sound intensity and alerts in case of excessive deviation (main systems thinking aspects: W, T, S, B).

4. Investigating the phenomenon of microphone feedback; exploring the relationship between the distance of the microphone from the loudspeaker and the frequency of the sound generated; discovering the sound wavelength; discovering how to prevent unwanted sound (main systems thinking aspects: W, I, F, D, S).
5. Investigating the process of adapting hearing aids to a person; addressing the advantages and limitations of a digital hearing aid compared to a traditional analogue device (main systems thinking aspects: W, T, S).

In conclusion, we have seen that sound systems are a rich technological field that is very close to a student's life and that exposes learners to all the principles of systems thinking we mentioned earlier in this chapter (Fig. 5.1).

Following the previous two chapters dealing with systems thinking in the context of the air conditioner and sound system, the next section will discuss key teaching and learning aspects related to fostering systems thinking in the science and technology class.

5.7 Fostering Systems Thinking in the Science and Technology Class: Key Teaching and Learning Characteristics

In this chapter, we address four key aspects of teaching and learning related to fostering systems thinking in the science and technology class, as shown in Fig. 5.22.

5.7.1 Modelling and Systems Thinking

Modelling relates to the generation of a physical, conceptual, graphical, or mathematical representation of a real phenomenon, process, concept, or operation that is difficult to observe or represent directly. Chan (2015) asserts that systems thinking is about building conceptual models that explain the complexity of the real world in terms of structure and behaviour. Scientific models explain and predict the behaviour of real objects or systems and are used in diverse scientific disciplines, including technology and engineering. Hallström and Schönborn (2019) assert that models and modelling are considered a fundamental aspect of STEM education, can be used to increase the relevance and authenticity of STEM disciplines, and contribute to the integration of STEM components in teaching and learning.

This chapter includes about 25 pictures, diagrams, drawings, and mathematical equations that represent different types of modelling, without which it would be almost impossible to present the aspects of systems thinking in electronics studies discussed here. However, creating or drawing a model of a physical, technological, or social system is not a simple task. Barak and Williams (2007) note that the process by

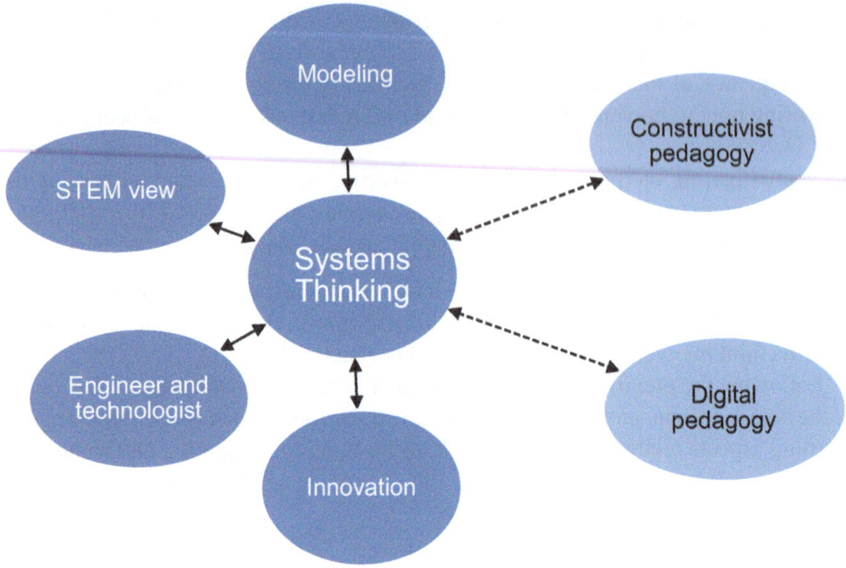

Fig. 5.22 Key characteristics of systems thinking and the instructional approach in the science and technology class

which people learn to use models or build their own models is a relatively unresolved issue in the psychological and educational literature. Therefore, it is important that the application of models to the teaching and learning of system structure and dynamics should be carried out gradually, with lots of examples from the simple to the complex. For example, the teacher can present in the class the block diagram of a heating device with a thermostat (Fig. 5.5) and ask the students to draw a block diagram of an air conditioner (Fig. 5.10).

5.7.2 STEM View and Systems Thinking

The term STEM—Science, Technology, Engineering, and Mathematics education—was proposed in the 1990s by the American National Science Foundation (NSF) and has been adopted widely by many countries around the globe. Sanders (2009) noted that although NSF has used STEM simply to refer to four separate and distinct fields, many educators and researchers have suggested that STEM education implies interactions between S, T, E, and M in the sense that 'the whole is greater than the sum of its parts'. On the other hand, STEM educators have established and consistently defended their sovereign territories (Kelley & Knowles, 2016). The two examples discussed in detail in this chapter—the air conditioner and the sound system—demonstrate the integration of STEM aspect in daily life. For example, learning that the air conditioner is based on concepts such as liquid evaporation and condensation and energy conversion, refers to science; concepts such as feedback, on/off control, and continuous control are about technology and engineering; concepts such as sound as a mechanical wave, frequency, and wavelength have to do with science; concepts such as electronic amplification and positive feedback relate technology, engineering, and mathematics. High school science and math courses should provide a sufficient foundation for STEM learning.

5.7.3 The Engineer, the Technologist, and Systems Thinking

Systems thinking is closely associated with the concept of engineering design, which usually includes the stages of identifying a problem or need, investigating, planning, constructing, evaluating, and improving. A major aspect of engineering design is the process of optimisation, which generally involves choosing the optimal solution from several options while balancing different requirements, for example, how to achieve high-quality outputs with minimum use of equipment, materials, or energy. Technology is a broader term than engineering, and engineering could be viewed as a branch of technology in which professionals use mathematics and scientific knowledge for the systematic design of products and systems. Kelley and Knowles (2016, p.6), based on De Vries (2011) and Barak (2012, p.138), write:

> Engineering can differ from technology in that engineering only comprises the profession of developing and producing technology, while the broader concept of technology also relates to the user dimension. Technologists, more than engineers, deal with human needs as well as economic, social, cultural or environmental aspects of problem solving and new product development.

Systems thinking is essential for both the engineer and the technologist since they must each understand how a system they are addressing works and its advantages and limitations. For example, engineers might face the challenge of developing an efficient, reliable, and relatively inexpensive air conditioner. The technologist deals with adjusting an air conditioner to an apartment, considering the customer's needs, the apartment size, the size of the windows and doors, the directions of the sun, and the typical weather in the place. An engineer engaged in sound system design is required to address several properties of the system, for example, the power of the amplifier and the loudspeakers, the system bandwidth, the quality of the microphone, mixer, amplifier, and loudspeakers. A technologist will deal with many factors associated with the users and the setting, for example, the type of music to be played, the audience, the environment, and how to avoid microphone feedback. Both an engineer and a technologist are required to address the seven aspects of systems thinking we discuss in this chapter: the whole rather than the parts, interconnections/interrelationships, feedback loops, dynamic behaviour, system boundaries, technological innovation, and socio-technological implications; however, the last three aspects might interest a technologist more than an engineer. Understanding the focus of an engineer versus a technologist may help educators develop systems thinking in the classroom.

5.7.4 *Innovation and Systems Thinking*

In the Introduction section of this chapter, we mentioned that we are currently at the beginning of the Fourth Industrial Revolution (IR 4), which is fundamentally changing the way we live, work, and relate to one another (Schwab, 2016). An interesting question is how new technological systems and devices are born. The traditional method for developing new products has been inspired by notions such as 'the voice of the customer' (VOC) or 'the voice of the market' (VOM). However, VOC and VOM may reflect only minutely customers' hidden wants and needs, especially in the case of a truly novel product, that is, a product that has never existed before. Abramov (2015) proposed supplementing VOC or VOM with the more objective 'voice of the product' (VOP), which shows that innovative products such as the iPhone were born from the knowledge, imagination, and initiative of the inventors and not due to a market survey. Innovative products are likely to create a market that did not exist before and often make the consumers say: 'how did we manage without that before'.

Systems thinking goes together with the innovation route mentioned above. To develop an innovative product or service, the inventor must have a broad view of the cultural, economic, and technological system for which the new product is intended. In our study, for example, it is natural that engineers would try to improve an air conditioner's technology, such as using inverter control technology. On the other hand, adding Wi-Fi to the air conditioner or an application that allows the consumer to operate the air conditioner via a smartphone is an innovative service that probably surprises and attracts customers. To what extent may this addition interest consumers and strengthen the product? How could we adapt the product to a wide audience? Dealing with such questions requires systems thinking related to the seven aspects discussed in this chapter by the company's designers and marketers.

5.7.5 *Instructional Approach for Fostering Systems Thinking*

5.7.5.1 The Role of Constructivist Pedagogy

Throughout this chapter, we have mentioned the advantage of constructivist pedagogy in the science and electronics class and presented some examples of how to apply these principles for fostering systems thinking in the context of an air conditioner and a sound system. However, educators are increasingly aware that at the beginning of learning a new subject direct instruction by the teacher might be more effective than constructivist pedagogy such as project-based learning because at these stages, the notion of minimal guidance during learning works only little (Barak, 2020). Some supporters of PBL (Hmelo-Silver, 2004; Savery, 2006) stress the need to tailor the scope and complexity level of assignments to students' prior knowledge and skills and provide instruction and support, to reduce the cognitive load and enable students to learn in a complex domain (Kirschner et al., 2006). By giving appropriate conceptual and representational scaffolding in the learning environment, students should be able to tap into their everyday experiences and channel and enhance these experiences to construct understandings of complex systems that are cognitively robust (Jacobson & Wilensky, 2006).

5.7.5.2 Using Digital Pedagogy

Educators and students now have diverse technological means at their disposal for teaching and learning technology and developing systems thinking. For example, in learning the subject of sound, students can use the Audacity free software to record and analyse the signals of different sounds. Figure 5.23 shows the electronic signal of two short beats in flute sounds recorded by a computer microphone.

Figure 5.24 shows the vibrations of the sound by zooming to a section of the signal seen in Fig. 5.23. From these two figures, students can learn to record and analyse analogy and digital signals, for example to identify the duration of each beat, the time difference between the two beats and the sound frequency. Students can accomplish this experiment with any laptop at school or at home, without the need for additional equipment. With appropriate instruction, the teacher can engage students in hands-on rich experiments, which are the key for learning all the seven aspects of systems thinking, for example, modelling, feedback loops, dynamic response, noise, and system boundaries.

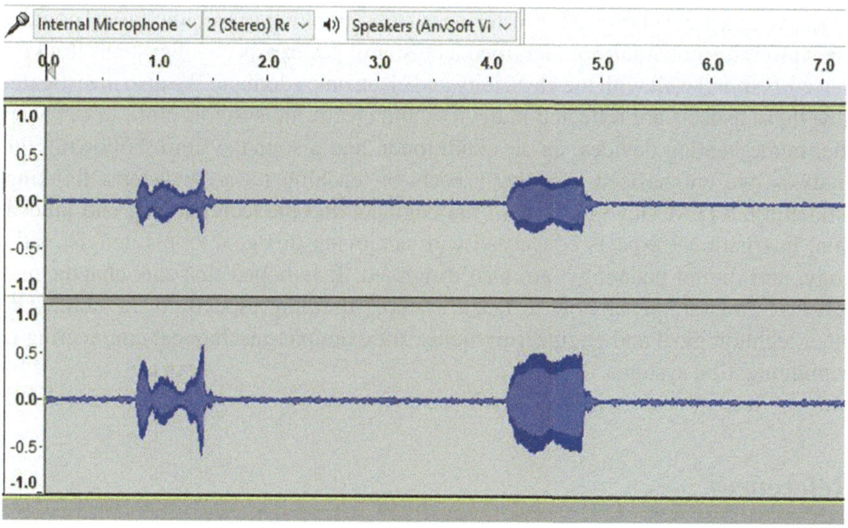

Fig. 5.23 A signal obtained from two short beats in a flute

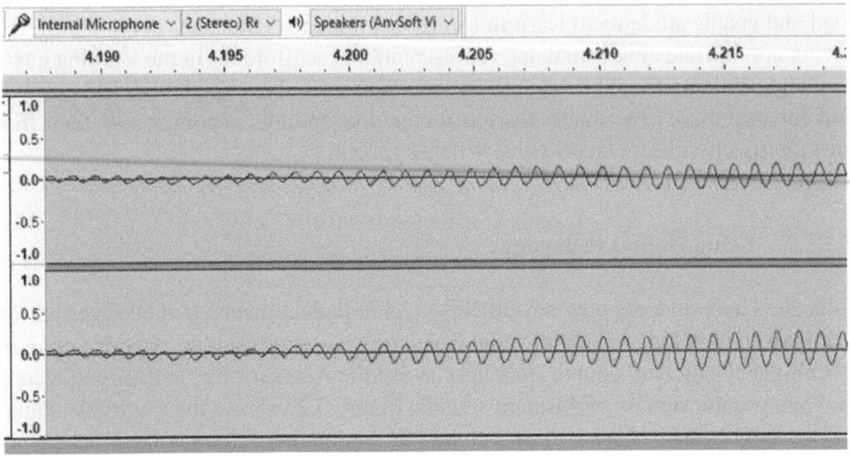

Fig. 5.24 Zooming into a section of the sound signal

5.8 Concluding Remarks

In this chapter, we have seen that systems thinking aims at understanding relationships between components in a system and their overall impact on a system's intended and unintended outcomes. This is an abstract, complex, and somewhat vague topic associated with terms such as higher-order thinking and experts' thinking. Electricity and electronics, which are at the heart of the latest technological revolutions, can serve as a good platform for fostering systems thinking in a technological context. To exploit this potential, we identified a cluster of seven aspects of systems thinking that correspond well with the electricity and electronics domain. We also investigated how these aspects are reflected in devices and electronic systems close to everyday life: home heating devices, an air conditioner, and a sound system. Following this analysis, we emphasised four key aspects of teaching towards systems thinking: modelling; STEM view; the role of the engineer and the technologist; and innovation. Instructional aspects concerned with achieving this goal, constructivist pedagogy, and digital pedagogy, are also discussed. It is hoped that this chapter may serve as an example of how to teach systems thinking aspects in the context of other technological and engineering fields, for example, mechanical engineering or communication systems.

References

Abramov, O. Y. (2015). 'Voice of the Product' to supplement 'Voice of the Customer.' In: *Proceedings of TRIZFest-2015 Conference, Seoul* (pp. 309–317). South Korea.

Amissah, M., Gannon, T., & Monat, J. (2020). What is systems thinking? expert perspectives from the WPI systems thinking colloquium of 2 October 2019. *Systems, 8*(1), 6.

Anderson, T. R., & Schönborn, K. J. (2008). Bridging the educational research-teaching practice gap: Conceptual understanding, part 1: The multifaceted nature of expert knowledge. *Biochemistry and Molecular Biology Education, 36*(4), 309–315.

Arnold, R. D., & Wade, J. P. (2015). A definition of systems thinking: A systems approach. *Procedia Computer Science, 44*, 669–678.

Barak, M., & Williams, P. (2007). Learning elemental structures and dynamic processes in technological systems: A cognitive framework. *International Journal of Technology and Design Education, 17*(3), 323–340.

Barak, M. (2012). Teaching engineering and technology: Cognitive, knowledge and problem-solving taxonomies. *Journal of Engineering, Design, and Technology, 11*(3), 316–333.

Barak, M. (2018). Teaching electronics: From building circuits to system thinking and programming. In M. De Vries (Ed.), *Handbook of technology education* (pp. 337–360). Springer.

Barak, M. (2020). Problem-, project- and design-based learning: Their relationship to teaching science, technology, and engineering in school. *Journal of Problem-Based Learning, 7*(2), 94–97.

Berg, R. E. (2020). Sound, the decibel scale. In: *Encyclopaedia britannica*. https://www.britannica.com/science/sound-physics/The-decibel-scale.

Bonta, D., & Snyder, S. (2009). New green home solutions: Renewable household energy and sustainable living. Layton, Utah, Gibbs Smith.

Briner, M. (1999). *Learning theories*. University of Colorado. Retrieved February 23, 2006.

Brown, J. L., Collins, A., & Duguid, P. (1989). Situated cognition and the culture of learning. *Educational Researcher, 18*(1), 32–42.

Bruner, J. S. (1961). The act of discovery. *Harvard Educational Review, 31*(1), 21–32.

Chan, W. T. (2015). The role of systems thinking in systems engineering, design, and management. *Civil Engineering Dimension, 17*(3), 126–132.

Chen, D. (2003). The system thinking in the PLC control system. *Control & Automation, 10.*

Daellenbach, H., McNickle, D., & Dye, S. (2012). Management science: decision-making through systems thinking. *Macmillan International Higher Education.*

De Vries, M. J. (2005). *Teaching about technology: An introduction to the philosophy of technology for non-philosophers*. Springer.

De Vries, M. J. (Ed.). (2011). *Positioning technology education in the curriculum*. Sense Publishers.

Dewey, J. (1929). *The quest for certainty*. Minton.

Forrester, J. W. (1994). System dynamics, systems thinking, and soft OR. *System Dynamics Review, 10*(2–3), 245–256.

Frank, M. (2002). Characteristics of engineering systems thinking-a 3D approach for curriculum content. *IEEE Transactions on Systems Man and Cybernetics, Part C—Applications and Reviews, 32*(3), 203–214.

Hallström, J., & Klasander, C. (2020). Making the invisible visible: pedagogies related to teaching and learning about technological systems. In: *Pedagogy for technology education in secondary schools* (pp. 65–82). Springer.

Hallström, J., & Schönborn, K. J. (2019). Models and modelling for authentic STEM education: Reinforcing the argument. *International Journal of STEM Education, 6*(1), 22.

Han, D. M., & Lim, J. H. (2010). Design and implementation of smart home energy management systems based on zigbee. *IEEE Transactions on Consumer Electronics, 56*(3), 1417–1425.

Hilton, M. (2010). *Exploring the intersection of science education and 21st century skills: A workshop summary*. National Academies Press.

Hmelo-Silver, C. E. (2004). Problem-based learning: What and how do students learn? *Educational Psychology Review, 16*(3), 235–266.

Holstein, W. K., & Bode, H. (2015). Systems engineering. In: *Encyclopaedia britannica*. https://www.britannica.com/topic/systems-engineering.

Hopper, M., & Stave, K. A. (2008). Assessing the effectiveness of systems thinking interventions in the classroom. In: 26th *International Conference of the System Dynamics Society* (pp. 1–26). Athens, Greece. http://citeseerx.ist.psu.edu/viewdoc/download?doi=10.1.1.422.1172&rep=rep1&type=pdf https://www.britannica.com/topic/systems-engineering.

Izushi, M. (2018). Energy saving through promotion of cooling. In: *The Japan Refrigeration and Air Conditioning Industry Association (JRAIA), Side Event at the 40th Session of the Open-ended Working Group of the Parties (**OEWG 40**)*, Vienna, Austria July 13. https://www.jraia.or.jp/english/side/presentation3_2018.pdf.

Jacobson, M., & Wilensky, U. (2006). Complex systems in education: Scientific and educational importance and implications for the learning sciences. *The Journal of the Learning Sciences, 15*(1), 11–34.

Kelley, T. R., & Knowles, J. G. (2016). A conceptual framework for integrated STEM education. *International Journal of STEM Education, 3*(1), 11. https://doi.org/10.1186/s40594-016-0046-z

Kirschner, P. A., Sweller, J., & Clark, R. E. (2006). Why minimal guidance during instruction does not work: An analysis of the failure of constructivist, discovery, problem-based, experiential, and inquiry-based teaching. *Educational Psychologist, 41*(2), 75–86.

Kopainsky, B., Alessi, S. M., & Davidsen, P. I. (2011). Measuring knowledge acquisition in dynamic decision-making tasks. In: *29th International Conference of the System Dynamics Society, Washington, DC* (pp. 1–31).

Lavi, R., & Dori, Y. J. (2019). Systems thinking of pre-and in-service science and engineering teachers. *International Journal of Science Education, 41*(2), 248–279.

McGraw-Hill Dictionary of Scientific & Technical Terms, 6E. (2003a). https://encyclopedia2.thefreedictionary.com/dynamic+behavior.

McGraw Hill Dictionary of Scientific & Technical Terms. (2003b). https://searchworks.stanford.edu/view/5177196.

Orgill, M., York, S., & MacKellar, J. (2019). Introduction to systems thinking for the chemistry education community. *Journal of Chemical Education, 96*(12), 2720–2729.

PhET Interactive Simulations. (2020). *University of Colorado Boulder* (open educational resources). https://phet.colorado.edu.

Pouspourika, K. (2019). The 4 Industrial Revolutions. Institute of Entrepreneurship Development. https://ied.eu/about/institute-of-entrepreneurship-development/.

Piaget, J. (1980). The psychogenesis of knowledge and its epistemological significance. In M. Piatelli-Palmarini (Ed.), *Language and learning* (pp. 23–34). Harvard University Press.

Richmond, B. (1994). Systems dynamics/systems thinking: let's just get on with it. In: *International Systems Dynamics Conference.* Sterling, Scotland.

Riess, W., & Mischo, C. (2010). Promoting systems thinking through biology lessons. *International Journal of Science Education, 32*(6), 705–725.

Sanders, M. (2009). STEM, STEM Education, STEMmania. *The Technology Teacher*, December/January, 20–26.

Savery, J. R. (2006). Overview of problem-based learning: Definitions and distinctions. *Interdisciplinary Journal of Problem-Based Learning, 1*(1), 9–20.

Schwab, K. (2016). The fourth industrial revolution: what it means, how to respond. *World Economic Forum, 14*(1).

Senge, P. (1990). *The fifth discipline, the art and practice of the learning organization.* Doubleday/Currency.

Squires, A., Wade, J., Dominick, P., & Gelosh, D. (2011). Building a competency taxonomy to guide experience acceleration of lead program systems engineers. In: *9th Annual Conference on Systems Engineering Research (CSER)* (pp. 1–10). Redondo Beach, CA

Sweeney, L. B., & Sterman, J. D. (2000). Bathtub dynamics: Initial results of a systems thinking inventory. *System Dynamics Review, 16*(4), 249–286.

Vygotsky, L. S. (1962). *Thought and language.* MIT Press.

Yakymenko, Y., Poplavko, Y., & Lavrysh, Y. (2020). STEM as a factor of individual systems thinking development for students of electronics speciality. *Advanced Education, 7*(15), 4–11.

Zuckerman, O., & Resnick, M. (2003). A physical interface for system dynamics simulation. In: *CHI'03 extended abstracts on Human Factors in Computing Systems* (pp. 810–811).

Moshe Barak is a Professor Emeritus in the Department of Science and Technology Education, Ben-Gurion University of the Negev, Israel, and formerly the Department Chair. His background is in electrical engineering, and he received his PhD in 1987 from the Department of Science and Technology Education at the Technion – Israel Institute of Technology. Barak has over 40 years of experience in teaching technology and engineering in high schools and colleges, the development and evaluation of innovative curricula and teacher training. His research interests focus on fostering higher-order cognitive skills such as problem solving and creativity in technology and engineering education and teaching advanced interdisciplinary technological subjects such as control systems and robotics to children. Barak has published over 140 papers in international refereed scientific journals, conferences and chapters in collective volumes.

Chapter 6
An Educational Model for Teaching About Technological Systems

Susanne Engström and Maria Svensson

Abstract Understanding and using models of systems are a part of systems thinking but there is a lack of functional educational models to use for this purpose. The aim of this chapter is to give an example of an educational model for teaching about technological systems to develop relevant and area specific systems knowledge and competences but also develop systems thinking as a key competence of technological literacy. Therefore, we want to take a theoretical model, the Freiburg model that describes system characteristics with the aim to develop systems thinking, and modify it to an educational model for teaching and learning about technological systems. Relevant knowledge and capabilities related to the common curriculum content in technology education in Sweden, the wastewater system, are implemented within the structure of the model with the intention to investigate if the theoretical Freiburg model is useful and understandable as an educational model. The answer is that teachers find the educational model helpful when modifying technological systems into activities for students in technology education. To become a useful educational model that provides all three aspects within technological systems teaching, the structural, the functional and the contextual, further development of the model is needed. However, we believe that this model is an important contribution to future studies of how an educational model could be a resource in technology education for developing students' systems thinking and systems knowledge.

Keywords Technological systems · Educational model · Functional aspects · Structural aspects · Contextual aspects

---

S. Engström (✉)
KTH Royal Institute of Technology, Gothenburg, Sweden
e-mail: sengstro@kth.se

M. Svensson
Göteborg University, Gothenburg, Sweden
e-mail: maria.svensson@ped.gu.se

J. Hallström and P. J. Williams (eds.), *Teaching and Learning about Technological Systems*, Contemporary Issues in Technology Education,
https://doi.org/10.1007/978-981-16-7719-9_6

6.1 From Theoretical Model to Educational Model

Technological systems are used to describe complex technical solutions often with a considerable number of components that interact with each other and with the surrounding context. Teaching about technological systems calls for developed knowledge of technical systems as well as systems thinking. The aim of this chapter is to give an example of an educational model for teaching about technological systems to develop relevant and area specific systems knowledge and competences but also develop systems thinking as a key competence of technological literacy.

In this chapter we illustrate the process of taking a theoretical model (Riess et al., 2015) that describes systems thinking and modifying it to an educational model for teaching and learning about technological systems within the systems thinking mode. We also exemplify the potential of such a model in technology education practice.

Theories provide an important scientific base in combination with proven experience when describing what constitutes teaching (see, e.g., Korthagen, 2005), but the question is how theoretical models become useful as an educational model in the school context. There are few contemporary educational models for technological systems and systems thinking. Therefore, it is important to investigate both how a theoretical model could be modified into an educational model and how this educational model could be a resource in technology education for developing students' knowledge.

As an example of a model for teaching about technological systems, related to both systems thinking and sustainable development (a mandatory and interdisciplinary perspective in a Swedish school context), we adopt the so-called "Freiburg heuristic competence model of systems thinking" (Riess et al., 2015, p. 18), and modify the theoretical frame to include technological systems and make it appropriate within the technology-teaching context. The educational model presents how technological systems knowledge can progress, from simple to complex, and from qualitative to quantitative aspects. The intended learning outcome for students is described through four different dimensions of systems thinking. The Freiburg model includes general knowledge of system concepts as well as the ability to create, use and evaluate system models. Understanding and using models (physical, symbolic, mental, mathematical, text-based) of systems are an integral part of systems thinking (see, e.g., Gillbert, 2004).

The concept of systems thinking has proven to be difficult to grasp for students and teachers. It includes understanding the whole as more than the sum of its parts, feedback, redundancy and more. It is a way of thinking and understanding that is central within technological system analysis. A technological system's interconnections provide the key to its function. A developed systems thinking allows us to identify interconnections and understand their relevance, essential for understanding problems and therefore in finding solutions. However, a deeper definition of systems thinking seems to be absent from the literature and there is also a lack of concrete models for how to educate about technological systems with the aim to develop systems thinking.

The Freiburg model mentioned above was developed to describe the knowledge and knowledge development in biology and geography, see Table 6.1. Other authors have also used it to analyse teaching about sustainable development in education, for example in Rosenkränzer et al. (2017). They stress that to develop systems knowledge it is important to understand the system aspects in relation to higher levels of systems thinking. Only those who reach these levels, which have to do with explanation and evaluation, can make in-depth analyses of systems.

In this chapter, we thus express how the theoretical Freiburg model provides a framework useful for modification into an educational model for technological

Table 6.1 Freiburg heuristic competence model of systems thinking (Riess et al., 2015, p. 18)

Competence dimensions	Sub-capability 1	Sub-capability 2	Sub-capability 3	Sub-capability 4
Dimension 4: Evaluation of system models	Determining the structural validity of system models	Determining the performance validity of system models	Determining the validity for the application	Determining the uncertainty of a forecast
Dimension 3: Solving problems using system models	Assessing the need for using a system model for processing the present problem	Assessing the type of system model (e.g. quantitative vs. qualitative) which is required to process a problem	Giving explanations, making predictions, and designing technologies based on qualitative system models	Giving explanations, making predictions, and designing technologies based on quantitative system models
Dimension 2: Modelling systems	Determining system elements, interactions, subsystems, system boundaries, system hierarchies and the model purpose	Understanding and reflecting on a complex system with the help of a text field or a word model	Reading and understanding qualitative system models. Constructing influence diagrams	Reading and constructing quantitative system models
Dimension 1: Declarative/conceptual systems knowledge	Basic knowledge of systems theory (system concept, system structure, system behaviour, sub-systems)	Knowledge of areas that can be considered as systems (also knowledge of simple and complex systems)	Knowledge of system hierarchies (e.g. cell, tissue, organ, organism, population, biocenosis, eco-system, biosphere)	Knowledge of properties of complex systems (structural and dynamic complexity, nonlinearity, emergence, etc.)

systems with the aim to develop systems thinking among teachers and students within technology education. The chapter includes some earlier research about technological systems, a description of the Freiburg model and how it may be modified into an educational model, then also some experiences that emerged among teachers trying out the educational model.

6.2 The Content of Technological Systems

During recent years, understanding of technological systems has become an essential part of technology education, and knowledge of what is required to develop students' understanding has increased. The argument is often that we need to understand how technological systems work in order to orient ourselves in a modern society. If students develop an understanding of the purpose, structure and behaviour of technological systems, this could facilitate the development of an ability to discuss and analyse complex issues in society (Klasander, 2010); especially issues related to sustainable development such as climate change, social equality and biodiversity, where technological systems impact solutions. However, students often have difficulty describing technological systems and the interaction between their components when the system is more or less invisible (Svensson, 2011). To support students' understanding of technological systems, teachers' awareness about what components to focus on, which connections they need to describe and how they contextualise the system, are essential.

One way to describe important components for understanding technological systems are as *functional, behavioural, structural* (see also Svensson's chapter) and *contextual aspects*. The functional aspects of technological systems focus on the understanding of the purpose of the system for humans and society (Klasander, 2010; Svensson, 2011; Svensson et al., 2012). The behavioural aspects concern the processes within components and how the input and output of matter, energy and information flow through the system; this enables learning of the system as a whole (de Vries, 2006; Koski & de Vries, 2013; Svensson, 2011). Concerning the understanding of structural aspects, there are internal structures, the physical parts which form the skeleton of the system, their functions and connections and overall descriptions of the system that facilitate the discovery of the different character of the relationship between components. There are also external structures such as relations between the system and systems and components in the surrounding context that are related to, but not directly part of, the system. These are described by Klasander (2010) as linear models, circular orbit models, hierarchical and network models. The contextual aspects are related to the interaction between the system and the surrounding environment, humans and other systems. To be able to distinguish the system from the rest of the world one important aspect is the system boundary (Koski & de Vries, 2013; Svensson, 2011). Identifying various systems and describing similarities and differences between systems is another important aspect of systems knowledge.

6.3 The Freiburg Model, Exemplified with a Technological System

In this section, we want to present an interpretation of the Freiburg model in relation to one technological system, wastewater. This technological system is often taken as an example in Swedish compulsory schools, for students aged 10–12. Our intention with this interpretation is to make the theoretical Freiburg model useful and understandable as an educational model. Relevant knowledge and capabilities related to the wastewater system are thereby implemented within the structure of the Freiburg model, see below in Table 6.2. In the same table, the four aspects, functional, behavioural, structural and contextual, found in earlier research, are integrated. Thereby, the aspects are related to the four competence dimensions in Table 6.1. According to Rosenkränzer et al. (2017), systems thinking is a multi-dimensional concept and they describe how knowledge within the area can progress, from simple to complex, from qualitative to quantitative, expressed in sub-capability 1 to 4 in Table 6.1. In sub-capability 4 compared with sub-capability 1, when going from left to right in the table for each dimension, more multi-faceted systems thinking is developed.

In its representation of the concept of systems thinking, the model includes general knowledge of system concepts as well as the ability to create, use and evaluate system models. Understanding and using models of systems are a part of systems thinking. Initially, in Table 6.1 the original theoretical Freiburg heuristic competence model of systems thinking is presented.

Table 6.2 Freiburg heuristic competence model of systems thinking. *Source* Riess et al., 2015, p. 18), interpreted by the authors for wastewater systems

Competence dimensions and related aspects to focus	Sub-capability 1	Sub-capability 2	Sub-capability 3	Sub-capability 4
Dimension 4: Evaluation of system model, comparison with wastewater systems in reality Functional, behavioural, structural and contextual aspects	Be able to see what aspects of the structure of the reality are depicted in the model Know which **parts and components** of the real wastewater system are represented in the model. Whether the model corresponds to reality in terms of structure	Be able to see how and to what extent the model depicts reality, in terms of **capacity and sequence**. Compare the wastewater model with the real wastewater system, in terms of size and dimensions, flows	Be able to evaluate the model's usefulness for **reality comprehension and problem solving** Whether the model can be used to simulate reality, discuss and solve any problems	Be able to evaluate the degree of consistency in the model's usability for **predictions.** Predictions regarding capacity, expansion, future problems, etc

(continued)

Table 6.2 (continued)

Competence dimensions and related aspects to focus	Sub-capability 1	Sub-capability 2	Sub-capability 3	Sub-capability 4
Dimension 3: Problem solving with the help of a model Functional and behavioural aspects, more of contextual and structural aspects	Be able **to start with a problem and realise the need for a system model** in order to be able to process the problem One example of a problem may be the risk of downtime in a wastewater system. If cotton swabs get stuck and stand upright Be able to reason about consequences and realise that a model can help one to understand	Be able to assess and evaluate **the type of model needed**, in terms of level of detail, size, and what quantitative and qualitative aspects to include Select parts of the wastewater system, which components are affected, how detailed the model needs to be. The type of model required. What the model needs to illustrate in order for the current problem to be understood and analysed	Be able to use a model to explain, draw conclusions from what has happened (regarding the problem) and what might happen (regarding the problem). Regarding **qualitative aspects** Based on the chosen model of the wastewater system, draw conclusions from how different things are affected by the problem, based on how the system works. Use the model to suggest preventive measures	Be able to use a model to explain, draw conclusions from what has happened (regarding the problem) and what might happen (regarding the problem). Regarding **quantitative** aspects Based on the chosen model of the wastewater system, draw conclusions from how the problem affects **the flow and is affected by the flow** in the system. Use the model to suggest preventive measures or improvements to the system
Dimension 2: Create a model of the system Focus on functional, behavioural and structural aspects, some contextual aspects	Be able to decide which parts of the system should be included in a model of the system. System boundary, various subsystems, components, etc. Be able to determine the purpose of the model Choose which components and interactions in a standard wastewater system are needed in a model	Be able to use a text-based model. After reading about the system, be able to describe the system's structure in relation to capacity, describe how water comes into the system, describe the system in plan and profile—how and why the water flows, what do the pipes look like, slope, etc	After reading and describing the system, be able to draw the system with its components and subsystems, be able to use appropriate symbols so that the drawn model describes how the system works	Be able to describe and create a quantitative model. Show how capacities in the system can be estimated. Describe a model that illustrates the capacity of the system: To understand the dimension-flow relationship. Be able to show how volumes and flows are affected

(continued)

Table 6.2 (continued)

Competence dimensions and related aspects to focus	Sub-capability 1	Sub-capability 2	Sub-capability 3	Sub-capability 4
Dimension 1: Knowledge of the system Functional, behavioural and structural aspects	Knowledge that **a system consists of interconnected parts** That the wastewater system consists of sewage pipes, manhole & wells, pumping stations, outlets, and a treatment plant (mechanical, biological, chemical purification) etc Know the actual concept of a system, a system structure, a system boundary, what is transported in the system, why it flows and that there are subsystems in the system	Knowledge of **the characteristics of** a system. Input-process-output. That a system can be different things That the toilet is a system, that the entire wastewater system is one, that the house's internal drainage system is one, that a pumping station is one, that part of a treatment plant is one, etc. There is an entrance, something happens, and there is an exit. In some systems it is easy to see what happens; in others it is more difficult	**Knowledge of system hierarchies** (Russian doll). That in one system there may be another system that is important for the system to work. That there is another system in *that* system For example, in the large wastewater system there is a treatment plant which is a system, in the treatment plant there is a mechanical treatment step which is a system, and in the mechanical step there is a pump that is a system	**Knowledge of the characteristics of complex systems.** Characteristics such as structural and dynamic complexity. That what happens in the system does not always happen linearly, that "accidents" happen, feedback occurs There may be characteristics of the system as a whole that are not present in individual components, or the nature of the system may change In the wastewater system, water flows for the purpose of transporting pollutants and specific waste The flow determines which additional quantities can be received. The slope and dimensions of the pipes determine the flow. Water comes flowing from various directions, water can be dammed up in wells

Schuler et al. (2018, p. 195) have used this competence structure table in a study of the teaching of systems thinking in relation to sustainable development. The authors stress that it is important to also reach the higher levels of Dimensions 3 and 4 of Table 6.1, which deal with explanation and evaluation—in order to be able to analyse and discuss issues in depth and in relation to the complexity and dynamics of the systems.

Table 6.2 presents the same model but with adapted texts about the wastewater system, interpreted and adapted to technology education in primary and lower-secondary schools.

6.4 The Transfer of the Freiburg Model to an Educational Model for the Wastewater System

The Freiburg model for systems thinking offers a structure for teaching about technological systems. By following the model, it is possible to both recognise and describe a technological system and learn to create and use a system model for structuring and analysing important elements and solving problems, and furthermore evaluate the system model in relation to the same system in the real world. This is an example of a modification from a theoretical competence model into a more concrete educational model useful for a teaching practice.

The educational model can be seen as a guide for teaching systems thinking in technology education. The guide consists of informative questions that a teacher can ask students, as well as examples of activities that students can carry out. The guide uses specific examples of the wastewater system. The guide will provide a basis for teaching the technological system, while also offering guidance on how it can be taught in a way that allows students to develop systems thinking according to a specific definition.

For each dimension and sub-capability, it is possible to formulate questions and create activities that make it possible for students to learn specific notions, develop a system model and use it for relevant analyses of a technological system. By going from sub-capability 1–4 the students progressively develop knowledge and abilities related to the dimension, with the aim to develop an understanding of a complex system. Therefore, in Table 6.2, there are examples of created questions and activities that can contribute to students' development of systems knowledge, taking into account their level of knowledge.

Content in the four dimensions and in the four sub-capabilities for each dimension are designed to develop students' abilities to recognise, describe and model (to structure, to organise) complex aspects of real systems. The students get the chance to develop the ability to identify important elements of the system and the varied interdependency between these elements. But also to practice the ability to recognise dimensions of time dynamics, and to construct a model of reality and to make prognoses and analyses on the basis of that model. By reasoning in such a way about what students will learn, we take inspiration from Riess and Mischo (2010) and their description of systems thinking.

In the descriptions below we start with Dimension 1 and the four sub-capabilities related to this dimension. Questions, short examples of informative texts and/or examples of activities are presented for each dimension and sub-capability.

Dimension 1. Knowledge of the system—Dimension 1 includes basic knowledge of a system, in this case the wastewater system. The student should be given the opportunity to develop this knowledge, in order to then move on and create, use and evaluate her/his own system model.

Relevant questions to emphasise in *sub-capability 1(express knowledge about structure and behaviour)*:

- *What are the parts of a wastewater system?*
- *What concepts are good to know?*
- *What is transported in the system?*
- *How does the transport occur?*
- *What are the different parts?*

Teachers should have knowledge about the system and be prepared to guide students about relevant notions, components, behaviour, purpose and structures.

Relevant questions to emphasise in *sub-capability 2 (express knowledge about the characteristics of the system)*:

- *What characterises a system?*
- *How can the toilet be described as a system? What controls the inflow?*
- *Wastewater is diverted through pipes running from different directions into a so-called "manhole well"; why and what happens in the well?*
- *Wastewater flows into a pumping station; what happens there?*

As a teacher it is relevant to have knowledge about how the large wastewater system consists of smaller subsystems, all with an input, a process and an output. It is important to describe the wastewater system as a whole, how wastewater containing faeces, toilet paper, cleaning agents/soap/washing-up liquid, etc. is piped in, in order to then be diverted and purified. How the wastewater after purification is discharged into streams, seas and lakes.

Relevant questions to emphasise in *sub-capability 3 (express knowledge about system hierarchies)*:

- *With focus on the treatment plant in the wastewater system, what parts does the treatment plant consist of?*
- *How can smaller systems be distinguished in slightly larger systems?*
- *What happens if any step in the treatment plant is inoperative?*

As a teacher it is appropriate to have knowledge about how all subsystems, for example, in the treatment plant, or a pumping station, have an impact on the whole.

Relevant questions to emphasise in *sub-capability 4 (express knowledge about the complexity within the system):*

- *How is the wastewater system connected?*
- *How is the wastewater flow connected? Is there a dynamic in the system?*
- *What happens in the large wastewater system if a well is clogged or a pump is inoperative?*
- *What happens to wastewater that cannot be diverted properly?*

As a teacher it is relevant to have knowledge about and have the ability to describe the system structurally and thereby express the complexity: About how the wastewater system is a system of pipelines that stretch across an entire city; it is in the wells that different pipes merge and are led further and then merge into other wells with other pipes; how the wastewater system is bounded by its pipes at the outer edge of the city; how the flow is determined by a combination of dimension and slope; how we can notice when the system does not work.

Dimension 2. Create a model of the system—In Dimension 2, the aim is to create a model of the wastewater system. Dimension 2 expresses abilities that students should be given the opportunity to develop. Based on the systems knowledge developed in Dimension 1, Dimension 2 thus focuses on creating one's own model, mental or physical, of the wastewater system.

Relevant questions to emphasise in *sub-capability 1 (be able to decide which parts of the system should be included in a model):*

- *Where does the wastewater system start; input? Where does it end?*
- *How can we delimit the system in the model?*
- *What components are important to include in the model?*
- *Why is the model being made? What do we want to understand? How do we want to use the model?*

Teachers should, for example, describe how the system boundary around the entire wastewater system is essential in a city (urban area). Teachers can choose to limit the discussion to a part of the system. The students should be guided to consider where there are inflows and outflows from the system and how the wastewater flows from one subsystem into another. As a teacher it is apposite to explain how a model can be created to illustrate how the system is structured and how different parts are connected within the system. In addition, how a model can be created to be able to understand important aspects such as the dimensions of the pipes and how and why the wastewater moves in the system. In such a case, it may also be interesting to specify the dimensions of the pipes so that one can understand how much wastewater can fit and flow in the system.

Relevant questions to emphasise in *sub-capability 2 (be able to use a text-based model to describe the system's structure and behaviour)*:

- *By reading about the wastewater system—what do you want to include in your model of the wastewater system?*

It may be important for the teacher to guide the students through text-based models of the system, found in literature or at homepages, to consider what components that could be important to visualise in their own model. Students may have to describe their understanding of the information about the system, how the water enters the system, how it is diverted, how the wastewater is passed through the system in order to end up in the treatment plant.

Relevant questions to emphasise in *sub-capability 3 (be able to draw the system with its components and subsystems, use appropriate symbols so that the drawn model describes how the system works)*:

- *Depending on what we want to use the model for, what should it look like?*
- *How can we get a picture of our own municipality's wastewater system?*
- *Depending on the purpose of the model, how should the system be delimited and what facts should be included?*
- *Description of the treatment plant? What subsystems? Various purification steps, volumes, quality of outgoing water?*

As a teacher it is relevant to have knowledge about the behaviour and the structure of the system, the pipe dimensions, the length of the pipes and the different subsystems. When the teacher guides the students in making a sketch of the parts chosen for their system model, a good idea is to get in contact with the municipality's wastewater management company or a water resource engineer who can show drawings of the entire system, and the subsystem parts.

Relevant questions to emphasise in *sub-capability 4* (*be able to describe a model that illustrates the capacity of the system):*

- *How are the dimensions of the pipes determined?*
- *What happens when one builds more buildings in the area?*

In order to be able to reason about the capacity of the system, the students may need a picture of the dimensions of the pipes and how long they are. They can also look at the depth of the pipes so they can estimate slopes. To do this the teacher should have relevant material, drawings and statistics from the municipality. It is impossible to attempt to consider an entire wastewater system in a municipality, but it may be interesting for the students to delimit one's examination to a part. Then the teacher can guide the students to estimate how other parts of the system work and what capacity they have.

Dimension 3. Using the model—problem solving with the help of a model—In Dimension 3, the model should be used by students to understand problems and to find appropriate solutions.

Relevant questions to emphasise in *sub-capability 1 (be able to start with a problem and realise the need for a system model in order to be able to process the problem)*:

- *What problems can arise in the wastewater system?*
- *How can the model help us to understand this?*

The teacher might have knowledge about relevant problems that can be solved by system models and how to guide students in using their own models to understand system problems. For example, grease plugs, plugged pipes due to things that have been flushed down, but also due to penetration by roots or breakage of the pipes. The model should provide possibilities to analyse the reasons for different problems. Knowledge about problems at the treatment plant is also relevant.

Relevant questions to emphasise in *sub-capability 2 (be able to assess and evaluate the type of model needed, in terms of level of detail, size, and what quantitative and qualitative aspects to include):*

- *If a specific problem is highlighted, what part of the model should be used and why?*
- *What does the model need to contain in order to analyse what is happening?*
- *For example, what happens upstream of the stoppage and with the flow?*

One teaching strategy is to guide the students to start from a common problem in the wastewater system. For example, an interruption in a pipe. It is relevant for students to determine where the problem occurred, consider which part of the model to use for an analysis. It is important to select pipes upstream from the interruption. The model should illustrate a clear demarcation of a subsystem, dimensions, pipe lengths, type of housing, levels of housing, the types of drains present in a dwelling, etc.

Relevant questions to emphasise in *sub-capability 3 (be able to use a model to explain, draw conclusions regarding qualitative aspects):*

- *What can happen when wastewater is pushed up into floor drains in homes and when nothing can be diverted from the washing machine or toilet?*
- *What problems arise for the people living in the home?*
- *How can one influence people to understand the system and think more what they should do?*

Knowledge about how the wastewater system interacts with people and their actions, are important to develop. The teacher needs to make students aware about what might happen if/when people affect the system through different actions. Moreover, they need to highlight what can be done to remedy the problem and to prevent it from happening in the future.

Relevant questions to emphasise in *sub-capability 4 (be able to use a model to explain, draw conclusions regarding quantitative aspects):*

- *What happens to the wastewater flow when a problem occurs?*
- *How much water can one estimate to fit in a pipe?*
- *How much is flushed out of a toilet? From a bathtub?*
- *How can we use the model to understand the consequences of problems in the system?*

The teacher may need to accentuate how to analyse the system capacity, what and why, when an interruption occurs in the flow and the wastewater continues to be diverted from homes and businesses. Why the wastewater starts to flow back and fill up into the floor drains at the bottom of the dwelling? It is worth noting that a pipe that diverts wastewater from a dwelling will soon become full if there is a stoppage further down the line.

Dimension 4. Evaluate one's own system model and compare it with real wastewater systems—Dimension 4 focuses on comparing the model created by the students with the real wastewater system. Of course, not everything can be included in one's own model but it is interesting to note what is different between the model and the system in reality, as well as what is the same. It can be complicated enough to transform a sketch or drawing into a three-dimensional model so as to compare this model with reality, but it is important for students to develop their own model after evaluating it against real-world conditions.

Relevant questions to emphasise in *sub-capability 1 (be able to determine the structural validity of system models)*:

- *What components exist in reality and which have been included in the model? Pipes, wells, pumping stations* etc.*?*
- *Does the model, for example, show the materials used for the pipes?*

In order for the student to be able to compare her/his own model with the real wastewater system, there must be drawings, pictures and films that show what it looks like in reality. A water resource engineer at the municipality who presents pictures, drawings, show movies and experiences can be an important asset in creating an understanding of how things happen in reality. If the teacher has chosen to focus on the treatment plant as a subsystem, a good idea is to do a field trip to the local treatment plant.

Relevant questions to emphasise in *sub-capability 2 (be able to compare the model with the real wastewater system, in terms of size, dimensions and flows):*

- *How large are the flows that are diverted?*
- *What is the length of the pipes that make up the municipality's wastewater system?*
- *How many pumping stations?*
- *What is the length of the stretches that flow using gravity?*
- *What aspects of what you have learned about the real system can be illustrated in the model?*
- *Can the model be further developed?*

As a teacher it may be appropriate to get knowledge about and the ability to illustrate flows and the capacity of various steps in the treatment plant, about the length of the pipes that make up the municipality's wastewater system and the size of the flows that are diverted.

Relevant questions to emphasise in *sub-capability 3 (be able to evaluate the model's usefulness, whether the model can be used to simulate reality, discuss and solve any problems):*

- *What are the most common problems in the real wastewater system?*
- *How much of the system has to be replaced each year?*
- *How does the municipality try to prevent problems from occurring?*
- *What aspects of what you have learned about the real system can be illustrated in the model, and how can the model help with analyses?*
- *Can the model be further developed?*

Students may need to have access to factual information that includes how the municipality use digital monitoring to control parts of the system, maintain broken pipes, how they clear them of penetrating roots, how they flush pipes and how they use a camera to inspect the insides of the pipes. There are facts, images and descriptions that are important for teachers to obtain, so that students can review it.

Relevant questions to emphasise in *sub-capability 4 (be able to evaluate the degree of consistency in the model's usability for predictions regarding capacity, expansion, future problems,* etc.*)*:

- *Which areas will be expanded/developed?*
- *What is the capacity of the wastewater system, is there capacity for more wastewater, or does a new subsystem need to be created?*
- *Where does the system have extra capacity and which parts of the system cannot handle the diversion of more wastewater? How does this fit with the areas where new housing is planned,* etc.*?*
- *What methods exist for increasing the capacity of an existing system?*
- *Perhaps there are opportunities to make a new wastewater network that connects to a new treatment plant?*
- *If one has chosen to focus on the treatment plant as a subsystem, one can ask about the capacities that exist in the various steps. Is there capacity for more wastewater?*
- *How can a wastewater treatment plant be expanded?*
- *Will there be new requirements for purification that require new methods?*
- *What aspects of what you have learned about the real system can be illustrated in the model, and how can the model help with analyses and predictions?*
- *Can the model be further developed?*

In order for the students to be able to compare her/his own model with the real wastewater system in terms of what might happen in the future, she/he needs facts about the real wastewater system and data on how the municipality plan to develop it. As far as the system of pipes is concerned, the teacher may get help from a water resource engineer at the municipality. They can tell about the plans in the municipality when it comes to the wastewater system, where they have extra capacity and which parts of the system that cannot handle the diversion of more wastewater.

6.5 Teacher Reflections on the Educational Model

To get an initial understanding of how the modification of the Freiburg model may work for teachers, a technology teacher tested it in two classes with students aged 11–12 and 13–16. The teacher's written and oral experiences and reflections on the educational model for the wastewater system constitute a first indication of the effectiveness of the model. Further investigations are needed for a more in-depth evaluation, but these first reflections help us to understand the potential of the educational model.

The educational model helped the teacher when transforming knowledge of the technological system into activities for students. Knowledge and abilities are presented in the educational model, and they correspond well with both the objectives of the technology syllabus and the subject content. The concepts mentioned in the educational model were interpreted by the teacher both as general concepts and as specific to the wastewater system. The teacher also stated that the indicative questions were relevant. The informational texts were considered instructive for the teacher but were also useful for the students. The collaboration with the municipal engineers and staff at the treatment plant proposed in the educational model was considered to be valuable and should work well especially if the contact is established beforehand.

However, the teacher expressed concerns about some difficulties such as the content that may be perceived as challenging for teachers. Many teachers lack experience and knowledge of technological systems in general and also lack knowledge of the specific wastewater system, despite the fact that they usually teach about it. Often the wastewater system is taught by describing the toilet and then a line is drawn to the treatment plant that is presented in more detail and a visit is made to the treatment plant in the municipality. The parts of the system between the toilet and the treatment plant are "blackboxed" and neither the structure nor the processes in between are made visible to the students. Either the students are asked to draw a system model with a toilet, pipes and a treatment plant, or they are asked to build a model with the same components.

The teacher also pointed out the omission of certain content in the model that is included in the course syllabus and which is desirable in a complete educational model. First, how and why the technological system has emerged and evolved over time. Another aspect that the teacher felt was lacking was a clearer focus on what requirements the system fulfils. The teacher also called for a concrete link to sustainable development in the educational model.

6.6 Learning Possibilities When Using the Educational Model

From the teacher reflections we interpret that the dimensions and sub-capabilities are helpful when structuring a teaching sequence and assessing students' reasoning about technological systems. The model points to relevant and concrete technology knowledge and relevant systems understanding. When the teacher used the model for assessment, it became clear that most of the students reached Dimension 4. The teacher adapted questions and activities to the students' levels of knowledge and relevant curriculum requirements that made it possible for the students to reach higher dimensions. Students created system models and made logical and realistic analyses both of volumes and functions of the wastewater system as well as of environmental issues connected to the wastewater system.

When using the educational model, the teacher also found it was possible to develop a systems knowledge progression where students aged 11–12 years old worked with Dimension 1 and 2, and then continued with Dimension 3 and 4 when they were 13–16 years of age. This would allow younger students to focus on developing their basic knowledge of the system and develop their own system model, in order to then use their system model for analysis and evaluation in the next step. However, for both learning possibilities and assessing aspects the educational model needs to be further investigated.

6.7 Discussion of the Educational Model

All these steps, dimensions and sub-capabilities, together seem to be relevant to systems thinking, according to Riess and Mischo (2010). Furthermore, we want to underline that the systems thinking model must be complemented with critical thinking. We want students to develop the ability to critically use system models; for example, understanding how technology is related to complex environmental problems. Therefore, it is relevant for students to develop their own system model and evaluate it against real-world conditions. This is important in order to increase the authenticity of the task, see, e.g., Hallström and Schönborn (2019).

It is however not always obvious that a technology teacher has the required systems knowledge, especially since there are many different more or less complicated systems in our society. Using the Freiburg model as an educational model thus fulfils a function by offering a clear structure and an opportunity to guide students to find information about systems. The model also has the potential to be used in teaching of different ages, starting with a less complicated system and thereafter adding more complicated and complex systems as students grow older.

The teacher whose reflection on the model was included above also pointed out an omission in the educational model: How and why technological systems have emerged and evolved over time. The original theoretical model (the Freiburg model) focuses solely on current systems and solutions of the future. However, the educational model can be complemented by retrospective questions. Another aspect is a clearer focus on the needs that the system meets. The fact that wastewater treatment meets human needs—the diversion of wastewater and subsequent purification of wastewater—is implicit at all stages and can be perceived as a general purpose of the system, but can probably be further clarified. The teacher also called for a concrete link to sustainable development. This aspect can also be seen as a general purpose of the system (to purify wastewater) but it can also be made more explicit.

Important for understanding technological systems are functional, structural and contextual aspects (see, e.g., Koski & de Vries, 2013; Svensson et al., 2012). In this study we have exemplified and investigated the use of a theoretical model as an educational model and it appears that the structural, behavioural and functional aspects of a technological system, the wastewater system, are clearly made visible in the educational model. However, the contextual aspects do not appear to be as obvious in the model. For students to develop an ability to discuss and analyse issues about sustainable development and solve environmental problems in society they need also to have an understanding of contextual aspects (Klasander, 2010). Also, the transferability of aspects when using the model for one system, such as the wastewater system, to another system, need to be investigated further. Therefore, the Freiburg model could be seen as a nascent educational model that has potential, after further development, to become a useful educational model that accommodates all aspects: the structural, the functional, the behavioural and the contextual. We believe that this model is an important contribution to future studies of how an educational model could be a resource in technology education for developing students' systems thinking and systems knowledge.

References

de Vries, M. J. (2006). Technological knowledge and artifacts: An analytical view. In J.R. Dakers (Ed.), *Defining technological literacy.* Palgrave Macmillan.

Gilbert, J. K. (2004). Models and modelling: Routes to more authentic science education. *International Journal of Science and Mathematics Education, 2*(2), 115–130.

Hallström, J., & Schönborn, K. J. (2019). Models and modelling for authentic STEM education: Reinforcing the argument. *IJ STEM Ed, 6*, 22. https://doi.org/10.1186/s40594-019-0178-

Klasander, C. (2010). *Talet om tekniska system: förväntningar, traditioner och skolverkligheter* (Doctoral dissertation, Linköping University Electronic Press).

Koski, M.-I., & de Vries, M. J. (2013). An exploratory study on how primary pupils approach systems. *International Journal of Technology and Design Education, 23*(4), 835–848.

Riess, W., & Mischo, C. (2010). Promoting systems thinking through biology lessons. *International Journal of Science Education, 32*, 705–725.

Riess, W., Schuler, S., & Hoersch, C. (2015). Wie laesst sich systemisches Denken vermitteln und foerdern? Theoretische Grundlagen und praktische Umsetzung am Beispiel eines Seminars fuer Lehramtsstudierende [How can systems thinking be taught and encouraged? Theoretical foundation and practical application in a course for teacher students]. *Geographie Aktuell Und Schule, 37*, 16–29.

Rosenkränzer, F., Hörsch, C., Schuler, S., & Riess, W. (2017). Student teachers' pedagogical content knowledge for teaching systems thinking: Effects of different interventions. *International Journal of Science Education, 39*(14), 1932–1951. https://doi.org/10.1080/09500693.2017.1362603

Schuler, S., Fanta, D., Rosenkraenzer, F., & Riess, W. (2018). Systems thinking within the scope of education for sustainable development (ESD)—A heuristic competence model as a basis for (science) teacher education. *Journal of Geography in Higher Education, 42*(2), 192–204. https://doi.org/10.1080/03098265.2017.1339264

Svensson, M. (2011). Att urskilja tekniska system. Didaktiska dimensioner i grundskolan. Ph.D. Thesis, Linköpings universitet, Norrköping.

Svensson, M., Zetterqvist, A., & Ingerman, Å. (2012). On young people's experience of systems in technology. *Design and Technology Education: An International Journal, 17*(1).

Susanne Engström is Associate Professor of Technical Science Education at KTH Royal Institute of Technology, Sweden, where she heads the Technology and Science Education research group. Her research mainly concerns attitudes to, gender aspects of and strategies in higher technical education, and teaching and learning about technological systems in compulsory school.

Example of one publication and one conference paper: Engström, S. (2018). Differences and similarities between female students and male students that succeed within higher technical education: profiles emerge through the use of cluster analysis. International Journal of Technology and Design Education, 28, 239–261; Engström, S., Norström, P., & Söderberg, H. (2021). A Model for Teaching Systems Thinking: A Tool for Analysing Technology Teachers' Conceptualising of Systems Thinking, and How it is Described in Technology Textbooks for Compulsory School. Techne serien—Forskning i slöjdpedagogik och slöjdvetenskap, 28(2), 241–251.

Maria Svensson is a researcher and senior lecturer in the subject area of Technology, University of Gothenburg, Sweden. She is also employed as senior guest researcher at Linnaeus University, Växjö, Sweden. Her research focus is in the field of technology education research with a special interest in technological systems and systems thinking. She also has an interest in technology teacher education and development of science and technology pedagogical knowledge. Two of her latest publications are: Svensson, M., Williams, P., von Otter, A. M., Larsson, J., & Sagar, H. (2021). Technology Content and Concepts in Preschool Teaching–A Practice-based Collaboration. *Techne serien-Forskning i slöjdpedagogik och slöjdvetenskap, 28*(2), 149–155 and Svensson, M. (2021). Systems in Everyday Lives: Making the Invisible Visible. In *Design-Based Concept Learning in Science and Technology Education* (pp. 192–203). Brill Sense.

Part III
Learning and Teaching About Technological Systems

Chapter 7
Young Students' Perceptions of Control Systems and Adaptive Robots

David Mioduser

Abstract In this chapter, we focus on young students' development of a systems worldview while learning-by-doing about control systems and adaptive robots. Artefacts capable of adaptive behaviours are part of our everyday environment at home, school, in- and outdoor environments, and work and leisure places. Young children encounter these artefacts in the form of, e.g., sophisticated toys, computer-controlled games and devices, kitchen appliances, elevators and automatic doors, or traffic light systems. This new "breed" of human-mind-made devices, that began populating our world only a few decades ago, is characterized by purposeful functioning capabilities, autonomous decision-making, programmability, knowledge accumulation capabilities and adaptive behaviour—challenging children's traditional and intuitive distinctions between the living and non-living realms; between inert and behaving agents; between the natural and the artificial. In the main sections of the chapter, following a brief presentation of the background for our work, we elaborate (based on evidence collected over the years in our studies) on three main issues: (1) young children's overall perception of adaptive systems; (2) their detailed understanding of structural, functional and behavioural features of these systems; (3) implications of our research insights for supporting children's learning of adaptive systems concepts and skills.

7.1 Introduction

Since the early years of the previous century, the development of systems theories and related methodological approaches marked a shift in perspective enabling scientists to study natural, social and technological phenomena focusing on aspects, interrelationships and processes that were traditionally overlooked. Theoretical developments (e.g., general systems theory, cybernetics, genetic algorithms, adaptive systems,

D. Mioduser (✉)
School of Education, Tel Aviv University, Tel Aviv-Yafo, Israel
e-mail: miodu@tauex.tau.ac.il

J. Hallström and P. J. Williams (eds.), *Teaching and Learning about Technological Systems*, Contemporary Issues in Technology Education,
https://doi.org/10.1007/978-981-16-7719-9_7

complex systems, or chaos; Simon, 1996) together with the development of sophisticated computer tools allowing the construction and study of models of systemic phenomena comprise a powerful intellectual toolbox affording scholars to represent complex systems, understand their structural and behavioural properties, and hypothesize about their emergent patterns of behaviour under changing conditions.

From an educational perspective, the integration of these approaches and tools into the curriculum is not a trivial endeavour. Neither is the design of appropriate pedagogical resources supporting teaching and learning from a systems worldview. Specific concepts characterizing complex systems (e.g., emergence, self-organization and nonlinearity) impose new ways of thinking and pose a serious learning challenge for many students. Moreover, this way of thinking and learning about the world stands in evident contrast with the rationale and pedagogical practices prevalent nowadays in formal educational systems.

In this chapter, we focus on young students' development of these novel worldviews and skills while learning-by-doing about control systems and adaptive robots. Artefacts capable of adaptive behaviours are part of our everyday environment at home, school, in- and outdoor environments, and work and leisure places. Young children encounter these artefacts in the form of, e.g., sophisticated toys, computer-controlled games and devices, kitchen appliances, elevators and automatic doors, or traffic-lights systems. This new "breed" of human-mind-made devices, that began populating our world only a few decades ago, is characterized by purposeful functioning capabilities, autonomous decision-making, programmability, knowledge accumulation capabilities, and adaptive behaviour—challenging children's traditional and intuitive distinctions between the living and non-living realms; between inert and behaving agents; between the natural and the artificial.

In the main sections of the chapter, following a brief presentation of the background for our work, we elaborate (based on evidence collected over the years in our studies) on three main issues: (1) young children's overall perception of adaptive systems; (2) their detailed understanding of structural, functional and behavioural features of these systems; (3) implications of our research insights for supporting children's learning of adaptive systems.

7.2 Background

The studies synthesized in this chapter were conducted within the framework of our long-term research program—over two decades of work with many colleagues and graduate students in our laboratory. Our main goals have been, and continue to be: (a) to unveil the multiple layers comprising young children's perceptions of control systems' nature and functioning; and (b) to trace closely children's gradual development of concepts and skills considered essential for understanding as well as for designing and programming these adaptive systems.

The rationale of our research program is grounded on three core bodies of knowledge: the characteristics of participant children's developmental stage; their understanding of the designed world in general, and control systems in particular; the pedagogical and methodological approaches required for scaffolding children's encounters with control systems concepts and features (guiding the development of the studies' intervention tasks).

The first relates to existing knowledge about the characteristics of the target age level (4– 7-year-old) that make it developmentally appropriate for the construction of key technological-thinking schemas and skills. The literature concerning developmental changes throughout the ages 5–7 pinpoint major transitions taking place at the cognitive, neuropsychological, emotional and social levels. Among these: changes in attentional capacity, working memory, executive functions and concepts acquisition (e.g., Case, 1992; Rueda et al., 2004); neuroanatomical or organic changes in the frontal and prefrontal areas of the brain, to which high cognitive functions are attributed (e.g., Anderson et al., 2001; Kanemura et al., 2003); changes in moral development, and in understanding the balance of positive and negative attributes (e.g., Case, 1992; Harter, 1996); and obviously the beginning of formal schooling and with it the socialization into a culture of structured patterns of learning. All these changes are indicative of a crucial transition stage at which conceptions, schemas, concepts, attitudes and intellectual tools evolve and become consolidated, forming the child's cognitive/affective stance towards the world—including the designed world.

The second foundational component in our rationale is the continuously evolving knowledge on children's perceptions of control systems. Research efforts have focused over the years on a range of issues, e.g., the overall stance adopted by children when reasoning about a system's behaviour either from a technological/engineering perspective or a psychological one (e.g., (Ackermann, 1991; Kuperman & Mioduser, 2012); affective and social issues in children's perception of robots (e.g., Turkle, 2017); or capability to plan and program a robotic system (Mioduser & Levy, 2010; Strawhacker & Bers, 2019). In the following sections, we will focus on specific aspects of this component while presenting the background of each theme. All our studies are in continuous interaction with this growing body of knowledge: our questions rely on it, and in turn, our findings contribute to its further development.

The third component relates to our approach for designing and scaffolding children's interaction with robotic systems in both real-world interventions and laboratory-based research settings (Mioduser et al., 2012; Spektor-Precel & Mioduser, 2015). The premises guiding the development of our research interventions and tasks are:

- Knowledge and understanding are gradually constructed as a result of meaningful interactions between existing knowledge and schemes and new triggering and challenging demands embedded in the task. In our studies, we aim to trace closely conceptual change and knowledge construction processes while children perform the tasks.
- The constructionist approach—the creation of a "public entity" (e.g., an artefact, a computer program and an explanation)—means the externalization and

objectification of inner (to the mind) thinking processes and knowledge. In this process, the learner's externalized thoughts are open to discussion, critical evaluation and insightful examination. Explanations, programs, visual representations, documented action—all these comprise the raw dataset examined in our studies.

A key research tool for developing the scenarios and task sets for each study is the KinderBot programming environment. Constructing the behaviour of the adaptive systems (i.e. programming) is a challenging task for the young children. For our studies, we have developed an iconic programming environment (which still continues to evolve) for planning, constructing and controlling the behaviours of a robot made of Lego, a programmable brick and sensors. The environment comprises a series of modes allowing programming in levels of increasing complexity: from the creation of sequential programs (representing one-time events), via packaging sequences of instructions in routines (reusable scripts) to the definition of a temporal and general rules using sensors' data to generate actions in changing conditions. Programming is based on the use of icons allowing intuitive and simple definition of commands without requiring writing or reading code. In Fig. 7.1, the screen for the more complex programming mode is presented: the definition of two rules (with underlying and/or/not functions) and usage of routines (the coloured boxes). Reference to the states of two types of sensors defines the condition part of the rules (in the figure touch and light sensors). The action cells hold either simple

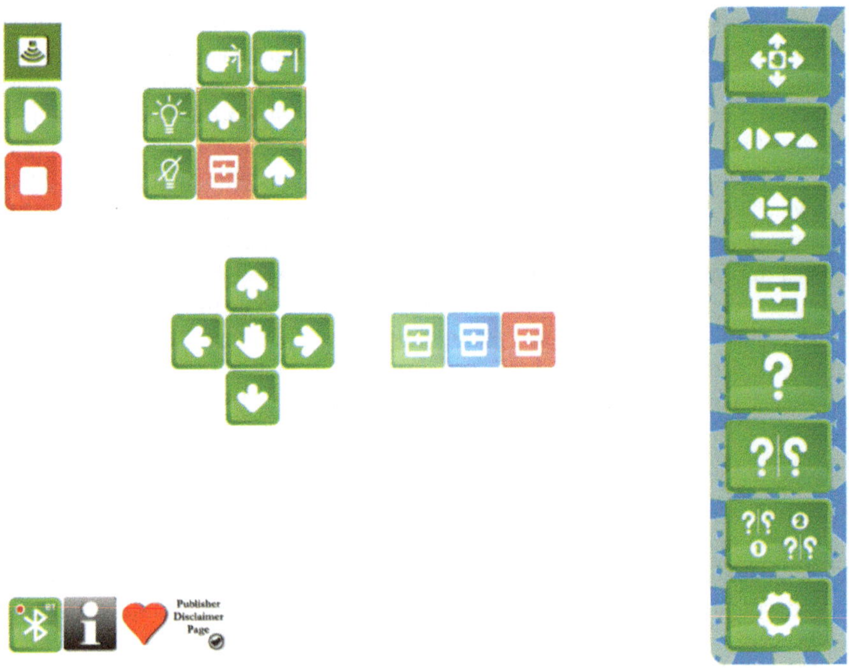

Fig. 7.1 Programming environment

commands for the robot's movement or a routine, a packed sequence of commands. In all our studies, children engage in actual programming tasks and are required to generate descriptions, explanations and representations of their work in tasks of increasing complexity. Sample tasks are: the robot has to navigate autonomously a space avoiding obstacles (low complexity), or to follow a path in the terrain without "falling" to either of the sides (complex task, using data from two sensors).

The generic methodological design of all our studies over the years includes the close examination of the work and explanations of children in either two modes of involvement in performing the tasks: as "observers-and-explainers" or as "programmers-then-explainers".

The integration of the above layers, namely the developmental characterization of the participant children, the evolving body of knowledge about their perceptions of control systems and the methodological premises of our research approach, is intertwined in the rational of our research program and each of the specific studies.

In the following, we will present segments of our observations concerning two main issues: Children's overall perception of the characteristics and features of adaptive systems, and their more specific perceptions of structural, functional and behavioural components of the systems.

7.3 Young Children's Perception of Adaptive Systems

7.3.1 Children's Mental Modelling of Technological Systems

7.3.1.1 Mental Models

Our mental model of a system is a core cognitive resource affecting the ways we interact with it—whether aiming to understand its structure and functioning, predict its behaviour in changing conditions and contexts, operate it, repair a fault in it or design a new one for fulfilling a similar goal or function.

The nature of a person's model of a device or a process has been described in varied forms. Kieras (1988) suggests that a mental model contains two forms of knowledge: (a) "How-it-works knowledge", referring to the internal structure and mechanisms of the device; and (b) "Strategic knowledge", about how to use the previous knowledge to perform a task. These two together result in a "runnable" mental model of the system.

Kleer and Brown (1983) suggest that constructing a model of a system implies: (a) a representation of the structural configuration of the system, called "device topology"; (b) a process by which the system's functional configuration is inferred from its structural components, called "envisioning"; and (c) a particular causal model resulting from the envisioning process. At the device topology level, the model comprises several constituents: **parts** (e.g., energy source, functional components),

conduits connecting the parts (e.g., pipes, wires) and "**stuff**" flowing through the conduits (e.g., oil, electrons, water, information).

Across diverse domains studied over the years, a common claim is that subjects' mental models affect their acquisition of knowledge and skills, and that the involvement in actual construction and modelling activities facilitates the development of a systems worldview.

Stemming from the research literature, the following claims are relevant for the issues presented in this section of the chapter:

- People use mental models to understand, explain, operate, repair or design a technological system.
- Mental models of a system are complex representations. They map the structure of the system (device topology), as well as the functions associated with these structural components, resulting in a "runnable" mental model of the system.
- Qualitative modelling precedes quantitative or formal modelling.
- People's previous naive knowledge and models are central elements in the modelling process.
- Experience (repeated activation of the model in varied situations, including construction and modelling activities) seems to play a central role in the gradual construction and refinement of the mental model.

Most studies have been conducted focusing on a user's device model while a person (usually an adult) is being trained to work with it. In these cases, the learning process is based on the need to acquire specific skills and strategies required by professional performance, e.g., operating a device or repairing it.

The studies presented in this section, in contrast, are concerned with a different population and a wider goal: (a) we focus on **young children's** mental modelling of technological systems; (b) we examine the more general context of **technological literacy**, defined as the knowledge and skills required for understanding and interacting with systems in the daily-life human-made environment; (c) we trace mental modelling while children are involved in **constructing** the adaptive systems' behaviour (programming tasks). Specifically, our focus is on children's mental modelling of simple **control** systems such as those found in automatic doors, heating–cooling systems and household devices. We explore children's conceptions and misconceptions, and the role of their evolving mental models for designing solutions in the realm of controlled and adaptive systems.

The main question leading our work on children's mental modelling of systems is: What takes for a child to go all the way from looking at a system as a whole in terms of Piaget's synthetic explanations stage (where there are no parts, no relationships among parts, mistakenly assigned functions to parts) up to grasping and understanding it as whole in systemic terms?

7.3.1.2 Mental Model of the Adaptive System

An adaptive system can be described by two main components: the operating unit (OU) and the control unit (CU), as depicted in Fig. 7.1 (Mioduser & Levin, 1996; Rodan, 2016). The OU comprises the configuration of physical components arranged in a way that elicits the fulfilment of the system's purpose or goal. The OU includes data collection components (e.g., sensors) and effectors of various kinds (e.g., motors, transmission compounds). The CU, by its conceptual definition, performs the processes required to generate appropriate outputs (e.g., instructions to activate a given effector) in correspondence with its program and incoming inputs (e.g., the reading of a distance sensor).

Norman (1983) suggested four standpoints in regards to the study of students' mental modelling of a technological system: (a) the target system [T]; (b) the conceptual model of the target system or [C[T]], such as its formal description by experts in a textbook or an operating manual; (c) the students' model of the target system or [S[C[T]]] their mental conceptual modelling of the system; and (d) the researcher's modelling of the students' models, or [R[S[C[T]]]]. Usually, the latest focuses on unveiling gaps between the [C[T]]—the experts agreed "correct" model—and [S[C[T]]] models, e.g., students' misconceptions and/or missing conceptions about any aspect of the system. Mostly, pedagogical plans and solutions aim to close observed gaps and support students' gradual mastery of the [C[T]].

7.3.1.3 Children's Mental Models of Adaptive Systems

Along our studies with young children (from kindergarten to 4th grade), we have examined the way they perceive adaptive systems and the changes in these perceptions as a result of their involvement in constructing the adaptive behaviours, i.e. in programming a robot. In the different studies, we used a wide range of adaptive systems, such as automatic doors, kitchen appliances with adaptive functioning, or simple robots built from Lego pieces.

The progression of children's mental models as observed in our studies and examples of children's explanations is depicted in Fig. 7.2 (Mioduser et al., 1996).

The progression evolves in four main stages from an initial undifferentiated input/output model up to a complete control model and is characterized by increasing differentiation of structural and functional components in the system and specialized comprehension of the role of these in the system's adaptive functioning. At first, the system is perceived as a black-box and referred to by its overall behaviour: in presence of an input (e.g., a person approaching an automatic door), it generates an output. This holistic view, which can be mistakenly viewed as "systemically holistic", corresponds to Piaget's synthetic explanations stage—perception of a whole without any differentiation among parts or functions.

Next in the progression is a "reactive" model. In all our studies with young children, sensing functions (data collection) and related physical components (e.g., a

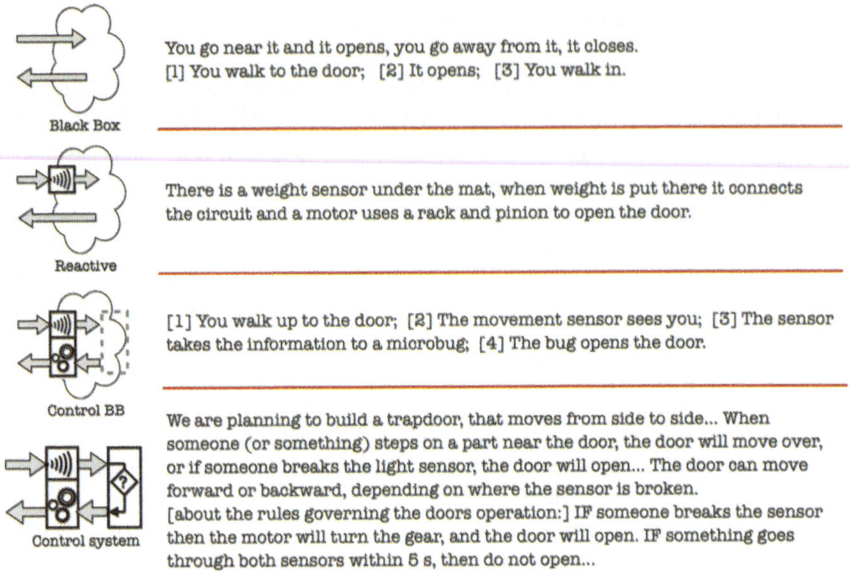

Fig. 7.2 Progression of mental models

light or touch sensor) are among the first structural components that they recognize and focus on, while working with an adaptive system. While children are able to allocate appropriate sensing functions and devices, a common misconception is that they endow the sensors with control capabilities (e.g., "it activates the motor"). Corresponding to the "implicit assumptions" type of misconceptions, children holding this model assume a direct effect of inputs to outputs without the need for a decision-making component.

Awareness for the need of such a function appears in children's "control black box" model of the system. A clearer differentiation of components appears, e.g., sensors, effectors, even a "mind" or a "microbug" (in a child's words) as well as the concept of flow of information within the system. Still it is not clear what the "insides" of the control unit are (e.g., rules, procedures, algorithms), but the conception of it is already there as well as the idea of some kind of tying between incoming information and outputs.

The complete "control model" completes the progression. This is a causal mental model in which control specifications, even formulated in some formal way, appear in children's explanations of the system and its functioning in the figure.

7.3.1.4 Qualities of children's Mental Models

In one of our studies, we evaluated the quality of children's mental models prior and following programming sessions and in relation to the formal model of the

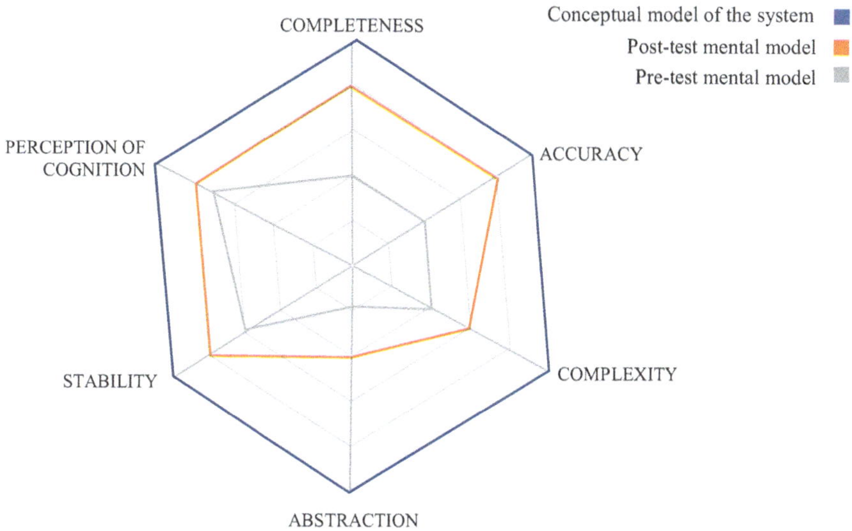

Fig. 7.3 Effect of programming on the quality of children's mental models

system (Rodan, 2016). Six criteria were defined: completeness, accuracy, complexity, abstraction and stability of the mental model, as well as perception of the system's "cognition" (Fig. 7.3). The findings indicated a substantial development of children's mental model of the robotic system: They developed a complete, precise, complex and quite abstract model after the programming sessions in comparison with the spontaneous models, even if still lacking some specific features of the formal (conceptual) model of the system. The change was not constant in all criteria: in the levels of completeness and accuracy of the mental model, a significant development took place, while in the level of complexity and abstraction there was moderate development. Mental model stability was maintained throughout the study, as well as the perception of the distinction between artificial and natural cognition.

From these findings, it can be assumed with a high level of certainty that the experience in programming the robot contributed to the improvement of children's cognitive schemas and complete mental models of the systems.

7.3.2 *Children's Stance Towards Technological Systems*

What conceptual perspectives guide children's thinking about autonomous/adaptive systems (e.g., robots)? Such behaving artefacts teeter precariously between the animate and inanimate world. Through their actions, and especially their intelligent behaviour, the inanimate world is brought closer to the living. Two distinct

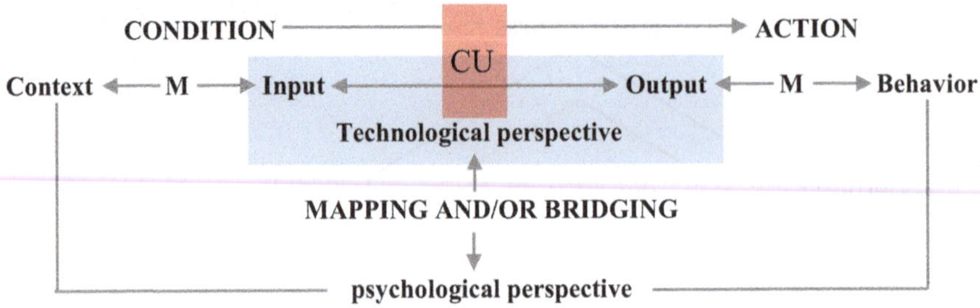

Fig. 7.4 Framework for analysing children's mental models

frameworks for approaching the world of behaving systems are suggested in the literature: the psychological and the technological stances (Ackermann, 1991). While the psychological stance attributes a robot's behaviours to human-like purposes, framed as animate intentions and emotions, personality and volition, the technological stance assigns causality to processes and mechanisms imbued in the inanimate material (i.e. physical parts such as motors and sensors) and the control program governing the system's functioning and adaptive behaviour. In our studies with young children, we examine these two perspectives in their perceptions of the systems.

The framework formulated by us for examining these perceptions appears in Fig. 7.4. The basic building block of the adaptive behaviour is the relationship between a given "condition" (e.g., data on environmental features such as the presence of an object to be picked or a slope in the terrain) and an "action" (e.g., a displacement in space or the activation of an effector) mediated by processing procedures in the control unit (CU). This basic building block is expressed differentially in each stance, i.e. as the interaction between contextual parameters and corresponding behaviours (psychological) or the interaction between inputs and outputs (technological). The gradual construction of a complete technological model of the system implies the mental mapping of contextual parameters into inputs, of observed behaviours into outputs, and of the holistic psychological perspective into a detailed technological perspective.

In many studies conducted by us over the years, we obtained conclusive evidence of the effect of young children's involvement in constructing the robot's behaviour on their gradual transition from psychological to technological mental models of the systems. A clear example is related to the language children use (terms, concepts and metaphors) while generating explanations of systems' features and functioning and of their work with these. We discuss this issue in the next section.

7.3.2.1 Use of Anthropomorphic Language

People's conception of the ambiguous status of adaptive systems has been a matter of study in previous work. In van Duuren and Scaife's study (1996), artefacts with different anthropomorphic features (e.g., adaptive behaviours) that can be interpreted by children as psychological reality, were used to elicit children's associations regarding issues such as mental acts of dreaming; motor acts of walking; sensory acts and feelings; and even the very question as to whether the objects have a brain. While children's ideas about a doll, book and person did not show any developmental differences, the "clever artefacts"—a robot and a computer—showed developmental differences.

Along similar lines, Francis and Mishra (2008) asked children (aged 3–8) to interact with "anthropomorphic toys" (e.g., a stuffed dog, a mechanical cat and a robotic dog). They asked children to interact with the toys and tell "if these are real". They report on differences between the children's verbal descriptions, mostly acknowledging that these are not real, and their behaviours, indicating confusion concerning the reality of, e.g., the robotic dog—the most sophisticated toy. In their study, they report on extensive use of anthropomorphic language as opposed to non-anthropomorphic language.

However, in most previous studies, the participants were requested to observe and/or interact with behaving artefacts—they were not involved in constructing their behaviour. As well, the focus has been on the use of anthropomorphic terms, and less attention has been put on the nature of non-anthropomorphic descriptions generated by the children, i.e. descriptions indicative of children's "intuitive engineering" (Pinker, 1997).

In our studies, we examined the effect of children's involvement in constructing a robot's behaviour on their thinking and perceptions. As mentioned above, in most of our studies, participants were divided into two groups: Observers (explaining-only) and constructors (programming-then-explaining) of the robots' behaviours.

Examples of children's explanations appear in Table 7.1.

Findings in a recent study conducted by us focusing on children's use of anthropomorphic language in their explanations show that (Kuperman & Mioduser, 2012):

Table 7.1 Types of explanations using different languages

Language	Foci	Explanation
Use of anthropomorphic language	Robot's behaviour is explained in terms of intentions, volition, feelings and human-like actions	"*… He sees that it is the sea and decides to turn…*" "*… If he sees a person then he has to tell him…*"
Use of technological language	Robot's behaviour is explained in terms of its components' functions, mechanisms and formal decision-making rules	"*… we simply wrote [programmed], when he gets to the black area he stops and when in the white area turns back…*" "*… and if one [sensor] sees white and the other sees black then [turn] left…*"

- The statements generated by the "observers" were similarly distributed concerning the use of anthropomorphic and technological language (50%), while the statements generated by the "constructors" were predominantly of the technological type (about two thirds).
- With the increase in tasks' complexity, the use of anthropomorphic language by the "observers" increased and by the "constructors" decreased. At the same time, the use of technological language by the "constructors" remained at a high level—about two thirds of the statements.
- For the more complex task, the "observers" generated a small number of statements, of these mostly using anthropomorphic language. For this task, the "constructors" generated the largest number of statements, mostly using technological language.

We obtained similar results in other studies conducted over time. In contrast with previous studies reported in the literature emphasizing kindergarten-age children's tendency to adopt animistic and psychological perspectives, we observed in our studies that the engagement in constructing the "anthropomorphic artefacts" behaviour promoted the use of technological language and indicated the early development of an engineering stance towards these systems.

We have observed that while approaching the "breed" of behaving and adaptive artefacts, children very rapidly adopt appropriate (even if not accurate or correct) language and explanatory approach. Epley et al. (2007) suggest that because social experience in early human life is primarily with human agents, understanding of non-human agents in non-anthropomorphic terms should develop later. A key condition for the development of a non-anthropomorphic stance is the parallel construction of alternative representations of non-human agents, resulting from increasing direct or indirect experience with these agents. We believe that the active interaction with robotic agents, up to the level of involvement in the construction of their behaviours as promoted in our studies, affords powerful opportunities for the development of alternative and non-anthropomorphic understanding of the features of non-human agents by the pre-schoolers.

7.4 Perception of the Structural, Functional and Behavioural Configuration of the System

In this section, we aim to dive into a detailed analysis of children's perceptions and understanding of specific structural, functional and behavioural (SFB) aspects of adaptive systems. As in all our studies, we examine children's perceptions prior and after their interaction with actual systems, either for explaining or for programming its behaviour.

7.4.1 *Perception of Structural and Functional Components*

In a recent study conducted with 50 kindergarten children (aged 5–6 years), we focused on their perception of the compound "operating unit" and the sensors configuration (Landsman, 2019). We examined children's perception of structural components in reference to the degree of familiarity of the system: a highly familiar system—an automatic door (as children encounter it in buildings, stores and elevators), and a less common system—an autonomous car following a path.

Although children perceive automatic doors as more ubiquitous and familiar systems, the **operating unit** and its components were perceived more correctly by most children in the autonomous car. It seems that the operating structural parts and functions in the car (e.g., wheels, transmission mechanisms and motor) are more evident for them than similar components in the automatic door (somehow hidden in its structure).

In contrast, a substantial component in any adaptive system, the sensing unit, was more correctly perceived by the children in the automatic door system. They correctly perceived not only the sensors, but also their location in the system's structure. By being a well-known system with a fairly evident goal, i.e. to open/close doors in response to input data (e.g., a person approaching the entrance), the core function "sensing" is obvious for the children. In contrast, the sensing function in the car was mostly misconceived, e.g., by its misplacement in the "headlights" perceived as "eyes". Thus, familiarity with a system, which indicates the existence of relevant schemas and models in the child's cognitive repertoire, implies also familiarity with sub-systems on the basis of their salient features and/or functions.

In the case of the car, existing schemas about **mechanical compounds** led to a detailed perception of components related to the generation and transmission of movement. In contrast, the less familiar (to the children) features of the CU in the car and the control process guiding its autonomous behaviour were more difficult to grasp.

7.4.2 *Representational Constructs Underlying a system's Behaviour*

Do children identify behavioural patterns underlying a system's behaviour? In our studies, we examined these questions using a framework that highlights the differences between three types of behavioural patterns or constructs: ***episode***, the least-general representation of behaviour, is a description of a unique series of steps which in sequential form generates the behaviour (e.g., the descriptive sequence: "the robot goes one step straight, turns right, six steps straight … goal reached"); ***script***, a generalized and temporally-sequenced representation, comprises sequential events packed in repeatable chunks or routines (e.g., a sequence of steps used by the robot for crossing **any** intersection with traffic lights, re-used as needed); ***rules***,

Table 7.2 Children's use of different representational constructs

		Group 1	Group 2	Group 3			
		Pre-school programmers	Pre-school explainers	First grade explainers	t 1–2	t 1–3	t 2–3
Task 1	Mean	2.86	2.02	1.97	9.03***	7.38***	0.35
	sd	0.26	0.37	0.52			
Task 2	Mean	2.93	2.63	1.94	3.55***	7.02***	4.44***
	sd	0.16	0.35	0.65			
Task 3	Mean	2.89	2.58	2.01	2.74**	5.76***	
	sd	0.21	0.48	0.71			

Representation structures scores: Score for episode = 1
Score for script = 2
Score for rule = 3
** $p < 0.01$ *** $p < 0.001$

the most abstract representation, are a temporal descriptions associating between environmental conditions and adaptive actions (e.g., condition-action or if–then-else compounds). In studies conducted by us over the years, we examined children's use of these representational constructs while they were engaged in either observing-and-explaining or programming-and-explaining—a simple robot's adaptive behaviour (Mioduser & Kuperman, 2020; Mioduser et al., 2009).

In Table 7.2, the findings of a recent study are presented. Sixty-nine children in three groups participated in the study: pre-school children "explainers", pre-school children "programmers" and first grade children "explainers". Data on children's use of the different constructs was collected along three tasks of increasing complexity by the demand of programming components: use of one rule, use of two rules and use of a rule and a routine (Mioduser & Kuperman, 2020).

Our findings showed that **5-year old's who program** the robot are able to grasp the complexity of the observed/expected behaviours of the robotic device and formulate it in the form of a temporal and general rules. We found that they do not use "episodes" at all to represent the behaviours and use "scripts" minimally. The use of rules among the **programmers** is dominant for all tasks administered—both easy and difficult. The difference between 5-year-old programmers and their peer explainers and between them and older children explainers (7-years-old) was significant for all tasks implemented.

Despite the developmental and maturation gap, by which we can expect a more robust rule-thinking by the older children, our findings (consistent over studies) indicate that this is not the case. In a previous work, we suggested two possible explanations for the older children's performance. The first relates to schooling—the acculturation process by which ways of thinking extensively supported by the flexible, experimentation-based and open-minded kindergarten's curricula and learning culture gets gradually replaced by the structured, academic-oriented and "right-processes" > > > "right-answers" curricula and learning culture in the school.

This forced thinking–framing culture makes difficult for the children to perform open-ended explorations of unstructured situations related to objects and systems in the world, and to generate appropriate insights and abstract explanations concerning their complex behaviour. The second explanation is related to the role of the active involvement in constructing the system's behaviour, which explainers only did not experience. We elaborate more on this issue in the final section on pedagogical issues.

About the rule-construction process by the children, in particular the 'programmers', it is marked by three features: increased generalization, a shift from temporal to a temporal constructs and decentring from the robot to include in the scheme both it and the environmental features affecting its behaviour. In the decentralization process, children abstract rules from the robot's behaviour by: (a) observing the robot's sequence of moves and actions in the landscape (episodes), with a primary focus on the robot's actions, rather than on the environmental conditions in a "robocentric" approach; (b) by seeking repeating routines in the robot's actions set off by particular features of the environment (scripts), partially decentring from the robot's actions; and (c) distilling a temporal relationships between the environmental conditions and the robot actions (rules), completing the decentring from the robot.

7.4.3 Perception of the Control unit—the "artificial Mind"

The core component of an autonomous adaptive system is the control unit—the configuration of components in charge of the intake of data, its processing and its translation into outputs to the operating sub-systems. By definition, the functional layer of the control unit is the essential ones to be grasped and understood—built-in in it is the logical configuration of interactions and processes which stand at the basis of the system's behaviours. Needless to say, this layer is an abstract formulation (compared with the underlying structural layer), and for this reason, it represents a serious challenge for young children. Nevertheless, we have seen in our studies how young children manage to perceive the control unit of the system. As an example, we will briefly refer to two studies conducted by us.

7.4.3.1 Perception of the Control unit's Structure and Functioning

In a study already mentioned (Landsman, 2019), we examined children's perception of a simple robotics system's control unit—its components and functioning for controlling the adaptive behaviour. First, we collected data about children's spontaneous perceptions, i.e. from their experience with control systems differing in degree of familiarity to them—automatic doors and autonomous cars. The second data collection point was after they had the opportunity to interact with the systems and even program a simple robot's adaptive behaviour.

Findings from the spontaneous stage show that children perceived more clearly the control unit and its functioning in the automatic door than in the autonomous car (the difference was statistically significant). Also concerning the marks for the perception of the behaviour as rule-based (i.e. the logical content of the unit), these were significantly higher for the automatic doors than for the car. A qualitative analysis of children's explanations showed a rule-like template underlying their description of the door's functioning. In contrast, given that children are less familiar with autonomous cars, they are not capable to construct appropriate modelling of its intelligence even if the basic building blocks for it are similar to these in the door.

Moreover, their spontaneous explanations about the interaction between the system and environmental conditions, the basis of the adaptive behaviour, showed a better understanding of it for the automatic door. In this case, the environment is "active": people approaching the door from both sides, aiming to enter or exit the elevator or store. The need for adaptive functioning in response to this interaction is clear. For the car, the environment "is simply there" (the road, an obstacle and a turn)—the interaction is more sophisticated and implies that the active role is played mainly by the car (both for data collection and generation of outputs).

Following the sessions in which children interacted with the robotic system (half of them observing-then-explaining and half of them programming-then-explaining the system's behaviours), the findings showed increase in the marks for most variables examined. However, as expected, children involved in programming tasks showed clearer and more complete perceptions of the control unit, its content and the system's autonomous functioning as a whole. Their explanations showed a clear departure from the concept that a person operates the system (as observed in the spontaneous stage). As well, they understood that a person is responsible of the system's autonomous behaviour by writing the control program.

7.4.3.2 Perception of the Artificial Mind

The second study relates to children's perception of the system's "artificial mind". The literature concerning human beings' conceptions of "traditional" artefacts is vast; however, little is known about our conceptions of behaving artefacts or of the influence of the interaction with such artefacts on cognitive development, especially among children. Since these artefacts are provided with an "artificial mind", it is of interest to assess whether and how children develop a cognitive construct to refer to its characteristics and functioning, a construct we named theory of artificial mind (ToAM).

Concerning the human mind, children's theory of mind (ToM) emerges at about the age of four or five. ToM refers to the ability to conceive mental states, knowing that other people know, want, feel or believe things which are different from our own (Premack & Woodruff, 1978; Gabriel et al., 2019). Children become aware that people have beliefs, not just about the world, but about the content of others' minds (e.g., about others' beliefs), and, like people's beliefs about the world, these too may be different or wrong. Most of the studies that focused on the pre-school child examined ToM by means of first- or second-order theory of mind tasks. First-order ToM involves reflecting on what someone is thinking or feeling. Second-order ToM involves predicting what one person thinks or feels about what another person is thinking or feeling (Westby & Robinson, 2014).

In a recent study, we aimed to examine the development of dedicates schemas by children concerning their perception of the "mind" governing adaptive systems' behaviours. We adopted the theoretical and methodological framework of ToM's research and expanded it by two additional layers. The first layer comprises an additional set of ToM tasks focusing on decision-making and problem-solving processes which were not part of the classic ToM set. Since the core of our enquiry is adaptive behaviour by artificial systems, this expansion was imperative. The second layer comprised a complete set of tasks based on the classic and expanded ToM tasks, but adapted to refer to the robot's adaptive processes and behaviour.

Examples of the new set of tasks appear in Table 7.3.

Table 7.3 Examples of ToAM tasks

Adaptation of classic ToM tasks to assess ToAM	In the "No will" task, the child is presented with a doll and a robot The child is told: "Roey has a robot. The robot can move forward or backwards. Roey programmed the robot to move forward. When he put the robot on the floor the robot moved backwards". Then, the child is asked the "no will" question, "What do you think happened?" or "Can this happen? If so—why? and if not—why not?". In addition, the child is asked to explain the answer To perform this task correctly, the child's answer should indicate that she/he understands that the robot's behaviour is the result of its programming and that it does not have a will, such as: "Roey didn't program the robot well" or "The robot was programmed to go backwards"
Tasks developed for the study to assess aspects that relate to behaviour control and adaptivity	In the "No will—decision-making task", the child is presented with a robot and a container filled with red and blue balloons The child is told: "When a child celebrates his or her birthday during the day red balloons are used. When the birthday celebration is at night blue balloons are used. There is a pile of balloons in the container. The robot is located next to the container". The child is asked the target question: "Which balloon will the robot pull out?" In addition, the child is asked to explain the answer To perform this task correctly, the child must say that if the robot was programmed to make decisions, it will first check the light outside (using light sensors) and thus will be able to pull out the correct balloons

In addition, as in all our studies, we addressed the effect of children's involvement in the construction of the "artificial mind" (programming the robot) on their ToM as well as on their schemas about the unit controlling the system's behaviour (ToAM).

The study was conducted with 24 children aged 5 and 7 (13 boys and 11 girls), data were collected along 12 sessions (ten of which for observation/programming tasks at four levels of complexity) using quantitative and in-depth qualitative methods.

A detailed presentation of the study's findings is obviously beyond the scope of this chapter (see Spektor-Precel & Mioduser, 2015). Taken together, our observations are indicative of the development of a particular construct characterizing children's understanding of the adaptive-control "mind" (ToAM), concerning the following:

- **First-order understanding**—children understood that:
  - Robots have no will; rather, their operation depends on either direct operation or programming.
  - Robots behave according to environmental conditions and make decisions consequently (rule-based behaviour understanding).
- Second-order understanding—children understood that:
  - A coordinated/interactive process in which robot "A" and robot "B" operate according to a program, and the operation of robot "B" depends on the outputs of robot "A".
  - The way robot "B" would behave following a change in the behaviour (output) of robot "A"—both dependent on their program.
- **Children's main model of the artificial mind**—They perceived the artificial mind as either: (1) ToM-like model—completely based on their model of the human mind; (2) ToM-based ToAM (TbToAM)—technological model referring to the artificial mind but using elements and terminology borrowed from ToM; (3) Partial ToAM model (PToAM)—technological model referring to components of the artificial mind (i.e. reference inputs and outputs of the robot) but not to processing processes (or algorithms) in the artificial mind; and (4) fully technological ToAM model (FToAM).
- **Flexibility of the artificial mind**—refers to the type of ToAM model in terms of generality: (1) "obeying" model—technological model referring to a mind following a pre-determined (linear/sequential) program suited to the task; (2) "adaptive"—technological model referring to a mind that makes decisions based on rules and environmental conditions.
- In a series of qualitative case studies with 5- and 7-years old children who either observed or programmed an adaptive robot, most of children's explanations reflected either a ToM-based ToAM or a full technological model. As well, older children showed better understanding of the adaptive character of the system's "mind".

7.4.4 *Perception of the system's Emergent Behaviour*

Emergence is regarded as key concept for understanding complex systems' behaviour (Chi et al., 2012). Basically, it demands understanding of macro-level patterns of behaviour and how these take form out of micro-level components' functions and interactions. The emergent macro-level behaviours "are-not-written" in any of the low-level components, but result from the dynamic interactions among these and between them and the environment. This highly abstract and complex concept become tangible, visible and graspable in the work with the programming environment (interface + physical robot). Young children have the opportunity to observe how a physical system's specific and temporal behaviours result from the interaction between underlying simple a temporal rules, data collected from the environment and the functioning of structural components. For example, in the task "robot dancing in a checkers-board", it navigates only on the dark squares—this overall behaviour is not expressly written in the very simple rule "*If on white then turn / If on black then go straight*" (such a rule is constructed using the interface's icons by the children). The emergent "dancing" behaviour and random jumps among dark squares result from the recurrent activation of the simple rule in correspondence with the inputs collected by the sensor (dark or white squares). In another task, "the guard of the island", the very same simple rule underlies a substantially different emergent behaviour—now the robot navigates a large black island instead of the checkers-board accordingly, the robot follows all the time the edge of the island, at the meeting between the black (the island) and white (the water) colours. These are two emergent behaviours resulting from the very same program, in response to different features in the environmental configuration. In these and other tasks, the emergent phenomena become tangible and concrete, as well as analysable and decomposable into its underlying low-level components—constructed and programmed by the child. This is the realm of concrete-abstractions for doing and thinking (Mioduser et al., 2009).

Focusing specifically on the definition of the a-temporal rules, a key building block in emergent behaviours, the findings in our studies show that pre-schoolers who program the robot are able to grasp the complexity of the observed/expected behaviours of the robotic device and formulate it in the rules form. Although many studies have shown that pre-school children have difficulties in coding and formalizing rules, our findings show a different picture: Children can use abstract tools to program a robot's actions and even explain its behaviour in terms of an abstract rule or even a compound of interacting rules and routines.

7.5 Pedagogical Principles—Supporting Young Children's Learning of Adaptive Systems

A substantial spinoff of our studies over the years are important insights integrated in our pedagogical rationale, methods and resources for supporting young children's learning of system thinking knowledge and skills. The rationale is built upon the following premises:

- Learner-centred approach: the learner is not only the main target of the pedagogical process, but also the main engine driving it.
- Knowledge is gradually constructed by the learner as a result of meaningful interactions between existing knowledge and schemes and new triggering and challenging demands embedded in the learning situation and a challenging learning environment.
- The constructionist approach—stressing the importance of the construction of a "public entity" for supporting the (inner to the mind) knowledge construction process.
- The sociocultural perspective: The view of knowledge as a social construct gives peer interaction and collaboration a substantial role in supporting the collective creation of knowledge.
- Learning by doing—to learn technology is to think technology and to do technology.

The specific expression of these premises for supporting the learning of technological systems and adaptive/autonomous systems guided us in the development of dedicated pedagogical methods and resources—for our studies and for our implementation programs for kindergartens as well. These are characterized by the following:

- Support for a **reflective** and **analytic** view of the world of technological systems. One typical task in our interventions is the "artefacts ID" assignment—the structural/functional/behavioural analysis of an artefact. Usually, children are requested to pick in their homes an interesting system (preferably these that are no longer in use representing previous solutions for a function fulfilled today by a current system). Aided by a member of the family, they are requested, using an ID template, to construct a structural and functional map of the system (including aspects such as materials they are made of, sub-systems, inputs and outputs, mechanical compounds or alternative solutions in other systems for the same goals and functions). Children's artefacts and maps are presented for plenary discussion in the kindergarten.
- **Participatory investigations**—active interaction with physical systems. A unique affordance of the physical robot is the possibility to accompany the functioning of the system through bodily interactions with the behaving object. For example, while planning navigation procedures, the child "plays the robot" or simulates its response to a light source. In a sense, both child and robot can be viewed as two agents in a multi-agent system. In such interactions, the child's role shifts from designer and observer to that of participant. This shift in roles affords multiple views of the system and its functioning.
- **Concrete-abstractions**. We saw how children's involvement in tasks affording activities at the same time symbolic (i.e. reflecting on the artefact's behaviour; working with the programming interface) and physical (i.e. manipulating and

observing the behaviour of a real artefact) supports their thinking and acting beyond the anticipated in the developmental literature for this age level. The possibility to move back and forth between the abstract and a temporal control rules on the programming interface, and their spatiotemporal enaction in the physical robot's behaviour, supports the creation of schemas connecting between the (symbolic) program and the (physical) system's emergent behaviour—otherwise a serious cognitive challenge for the young children.

- **The programming environment**. For our studies, we have developed the KinderBot iconic programming environment (which still continues to evolve). All previous pedagogical premises are afforded by the work with KinderBot, which has proved to be a powerful tool supporting children's construction of knowledge.
- **Children's Zone of Proximal development**. Undoubtedly the presence of the physical construction and programming systems in children's ZPD afforded their coping with challenging tasks. But obviously the presence of the "knowledgeable other"—either the researcher or the teacher, supporting children's explorations, played important role as we could see in our studies. We have observed the children grew in their understanding of central concepts related to cybernetics: principles of feedback and control, emergent patterns that result from interactions between multiple rules underlying the robot's operation, its physical structure and its environment. When further support (in the constructionist way) was present, their abilities were augmented. They appropriated the offered tools, thinking in "concrete-abstractions" about the robot's behaviour and generating impressive solutions for the tasks.

7.6 Concluding Remarks

This chapter presented selected segments (by no means an exhaustive summary) of our long-term research program aiming to unveil the different layers of young children's understanding of controlled adaptive systems. Common to all our studies is the methodological setting in which children are engaged in constructing the systems' adaptive behaviours (i.e. programming a robot to perform tasks of increasing complexity).

Our studies' findings indicate that the young "constructors of behaviours" developed a profound and complex understanding of the adaptive systems along and as result-of their programming experiences. Following the programming sessions children developed complete, precise, complex and quite abstract mental models of the adaptive system—from spontaneous and undifferentiated models up to complete control models with differentiated control features and functions.

In contrast with previous studies emphasising kindergarten children's tendency to adopt animistic and psychological perspectives, we observed in our studies the early development of the use of technological (non-anthropomorphic) language as well as an engineering stance towards the adaptive systems.

We observed a steady development in children's understanding of the structural (e.g., components, compounds), functional (e.g., flow of information) and behavioural (e.g., goal-oriented, emergent behaviours) layers of the systems. We observed as well the development in children's capability to create appropriate representations of the programs required to control the system's adaptive behaviour (from linear event-like descriptions up to a temporal rules). And above all these, we obtained evidence of children's gradual development of the overall perception of the features of the "artificial mind" governing the functioning of the adaptive systems.

Our research agenda is still developing. Building on the work already done, we are about to explore several paths aiming to add new knowledge-segments to our understanding of children's thinking about intelligent systems.

Acknowledgments The work reported in this chapter is by no means a solo endeavour. Numerous colleagues and students are equal partners in the long-term research program conducted (and still being conducted) over the years. Participating research associates in different segments: Dr. Asi Kuperman, Dr. Gonen Rave, Prof. Sharona T. Levi, Dr. Michal Levi, Dr. Karen Spektor-Precel, Dr. Osnat Dagan, Inbar Rodan, Limor Landsman, Michal Biran and last but not least all our partner kindergarten teachers and children. Segments of the work were partially funded by the Israeli Academy of Sciences and TAU's Science and Technology Education Center grants.

References

Ackermann, E. K. (1991). The "Agency" model of transactions: Toward an understanding of children's theory of control. In J. Montangero & A. Tryphon (Eds.), *Psychologie Genetique et Sciences Cognitives* (pp. 63–73). Foundation Archives Jean Piaget.

Anderson, V., Northam, E., Hendy, J., & Wrennall, J. (2001). *Developmental neuropsychology: A clinical approach.* Psychology Press.

Case, R. (1992). The role of frontal lobes in the regulation of cognitive development. *Brain and Cognition, 20*, 51–73.

Chi, M. T. H., Roscoe, R. D., Slotta, J. D., Roy, M., & Chase, C. C. (2012). Misconceived causal explanations for emergent processes. *Cognitive Science, 36*(1), 1–61. https://doi.org/10.1111/j.1551-6709.2011.01207.x

de Kleer, J., & Brown, J. S. (1983). Assumptions and ambiguities in mechanistic mental models. In D. Gentner & A. L. Stevens (Eds.), *Mental models.* Erlbaum.

Epley, N., Waytz, A., & Cacioppo, J. T. (2007). On seeing human: A three-factor theory of anthropomorphism. *Psychological Review, 114*(4), 864–886.

Francis, A., & Mishra, P. (2008). Differences in children's verbal responses and behavioral interactions with anthropomorphic artifacts. In *Paper presented in the AERA 08 Annual Meeting.* New York, NY.

Gabriel, E., Oberger, R., Schmoeger, M., Deckert, M., Vockh, S., Auff, E., & Willinger, U. (2019). Cognitive and affective theory of mind in adolescence: Developmental aspects and associated neuropsychological variables. *Psychological Research Psychologische Forschung, 85*, 533–553. https://doi.org/10.1007/s00426-019-01263-6

Harter, S. (1996). Developmental changes in self-understanding across the 5 to 7 shift. In A. Sammeroff & M. Haith (Eds.), *The five to seven year shift* (pp. 207–236). Chicago, IL: The University of Chicago Press.

Kanemura, H., Aihara, M., Aoki, S., Araki, T., & Nakazawa, S. (2003). Development of the prefrontal lobe in infants and children: A three-dimensional magnetic resonance volumetric study. *Brain & Development, 25*, 195–199. https://doi.org/10.1016/S0387-7604(02)00214-0

Kieras, D. (1988). What mental model should be taught: Choosing instructional content for complex engineered systems. In J. Psotka, L. D. Massey, & S. A. Mutter (Eds.), *Intelligent tutoring systems—lessons learned.* Erlbaum.

Kuperman, A., & Mioduser, D. (2012). Kindergarten Children's Perceptions of "Anthropomorphic Artifacts" with Adaptive Behavior. *Interdisciplinary Journal of E-Skills and Lifelong Learning, 8*, 137–147. https://doi.org/10.28945/1732

Landsman, L. (2019). *Preschool children perceptions of how control systems work following daily usage and following working experience with a control system. Unpublished M.A. Thesis, (Hebrew).* Tel Aviv University.

Mioduser, D., Kupereman, A., & Levy, S. T. (2012). Design and learning (D & L) in the kindergarten. *Constructionism, 2012*, 620–624.

Mioduser, D., & Kuperman, A. (2020). Young Children's Representational Structures of Robots' Behaviors. *Design & Technology Education, 25*(2), 143–159. http://search.ebscohost.com/login.aspx?direct=true&db=bri&AN=144914633&site=ehost-live&authtype=ip,uid.

Mioduser, D., & Levin, I. (1996). Cognitive-conceptual model for integrating robotics and control into the curriculum. *Computer Science Education, 7*(2), 199–210. https://doi.org/10.1080/0899340960070205

Mioduser, D., & Levy, S. T. (2010). Making sense by building sense: Kindergarten children's construction and understanding of adaptive robot behaviors. *International Journal of Computers for Mathematical Learning, 15*(2), 99–127. https://doi.org/10.1007/s10758-010-9163-9

Mioduser, D., Levy, S. T., & Talis, V. (2009). Episodes to scripts to rules: Concrete-abstractions in kindergarten children's explanations of a robot's behavior. *International Journal of Technology and Design Education, 19*(1), 15–36. https://doi.org/10.1007/s10798-007-9040-6

Mioduser, D., Venezky, R. L., & Gong, B. (1996). Students' perceptions and designs of simple control systems. *Computers in Human Behavior, 12*(3), 363–388. https://doi.org/10.1016/0747-5632(96)00014-3

Norman, D. (1983). 'Some Observations in Mental Models'. In D. Gentner & A. L. Stevens (eds.), *Mental Models.* van Duuren & Scaife

Pinker, S. (1997). *How the mind works.* W. W. Norton & Co.

Premack, D., & Woodruff, G. (1978). Does the chimpanzee have a 'theory of mind'? *Behavioral and Brain Sciences, 4*, 515–526.

Rodan, I., (2016). *Development of mental models of control systems by 7 years old children engaged in robot programmimg. Unpublished M.A. Thesis, (Hebrew).* Tel Aviv University.

Rueda, M. R., Fan, J., McCandliss, B. D., Halparin, J. D., Gruber, D. B., Lercari, L. P., & Posner, M. I. (2004). Developmenat of attentional networks in childhood. *Neuropsychologia, 42*, 1029–1040. https://doi.org/10.1016/j.neuropsychologia.2003.12.012

Simon, H. (1996). *The sciences of the artificial* (3rd ed.). Cambridge, MA: M.I.T. Press.

Spektor-Precel, K., & Mioduser, D. (2015). The influence of constructing robot's behavior on the development of Theory of Mind (ToM) and Theory of Artificial Mind (ToAM) in young children. *Interdisciplinary Journal of E-Skills and Lifelong Learning, 11*, 329–345. https://doi.org/10.1145/2771839.2771904

Strawhacker, A., & Bers, M. U. (2019). What they learn when they learn coding: investigating cognitive domains and computer programming knowledge in young children. In *Educational technology research and development* (Vol. 67, No. 3). Springer US. https://doi.org/10.1007/s11423-018-9622-x

Turkle, S. (2017). *Alone together—why we expect more from technology and less from each other.* Basic Books.

Van Duuren, M., & Scaife, M. (1996). "A robot's brain controls itself, because a brain hasn't got a brain, so it just controls itself"—Young children's attributions of brain related behaviour to intelligent artefacts. *European Journal of Psychology of Education, 11*, 365–376.

Westby, C., & Robinson, L. (2014). A developmental perspective for promoting theory of mind. *Topics in Language Disorders, 34*(4), 362–382. https://doi.org/10.1097/TLD.0000000000000035

David Mioduser is PhD and Professor Emeritus of Science and Technology Education at Tel Aviv University. He served as Dean of the Jaime and Joan Constantiner School of Education, head of the Science and Technology Education Center, and head of the Department of Mathematics, Sciences and Technology Education in the School of Education. His research work addresses: (a) cognitive and learning issues in technology education, with specific focus on young children's learning - his work examines the development of children's conceptions and their construction of knowledge and skills concerning the understanding and making of the human-mind-made world; and (b) the meaningful incorporation of advanced technologies into teaching and learning processes. Current work focuses on: Learning by Modelling complex systems; robotics and control systems in education. He has been a research partner in international studies conducted under the auspices of the European Union, the OECD and the IEA, and member of leading policy-making committees dealing with ICT in education and Technology Education.

Chapter 8
Feedback in Technological Systems

Jonas Hallström

Abstract Feedback mechanisms make control of systems automatic and are thus inherent features of many technologies that surround us in our daily lives. Feedback is thus considered important to understand in technology education, although it is regarded as difficult and often not introduced to students until upper secondary level. Given the central role of feedback in technology and engineering, it is surprising that there is virtually no research on how students of any age conceive of and/or learn about feedback in the technology and engineering education literature. The aim of this chapter is to report a two-cycle intervention to improve Swedish secondary pre-service technology student teachers' conceptions of feedback in technological systems and to generalize some possible suggestions based on this study. Eleven student teachers altogether took part in the two cycles of the intervention, taking a pre-test prior to it and a post-test afterwards. Although this is a small sample, overall the findings indicate that the student group as a whole performed slightly better in the post-test than in the pre-test, which was particularly obvious in cycle 1. In cycle 2, the students did not perform quite as well in the post-test as in the pre-test, despite an improved intervention based on the findings in cycle 1. The findings also suggest that some teachers understood the systemic aspects of feedback mechanisms better after the intervention. On the other hand, no student reached an expanded understanding, and most conceptions were rather vague. Furthermore, there was a general lack of atomistic conceptions, for example, sensors and how they work in a control system. This study thus confirms previous research about the lack of essential device knowledge among student teachers.

Keywords Student conception · Control theory · Feedback mechanism · Technological system · Technology teacher education

---

J. Hallström (✉)
Department of Behavioural Sciences and Learning, Linköping University, S-60174 Norrköping, Sweden
e-mail: jonas.hallstrom@liu.se

J. Hallström and P. J. Williams (eds.), *Teaching and Learning about Technological Systems*, Contemporary Issues in Technology Education,
https://doi.org/10.1007/978-981-16-7719-9_8

8.1 Introduction

Modern control theory was developed in conjunction with the emergence in the twentieth century of increasingly complex technological systems such as urban systems, weapons systems and communications systems (Hughes, 2004; Mindell, 2002). In essence, control theory is therefore about controlling artefacts and systems and making them behave in the way we want them to (Glad & Ljung, 2000, 2006), by employing controllers and sensors for primarily negative feedback to ensure the reliability, predictability and stability of the systems (see Appendix). Feedback mechanisms make control of systems automatic and are thus inherent features of many technologies that surround us in our daily lives, from the home heating system to the municipal sewer system to automatic flight control. Feedback is considered crucial to learn in technology education (Barak, 2018; Hacker, 2018), although it is regarded as difficult and often not introduced to students until they reach upper secondary level (e.g. Goovaerts et al., 2019; Martin, 1990). Some technology curriculum documents and standards for K-12 education thus include feedback (e.g. *Standards for Technological Literacy: Content for the Study of Technology*, 2007), whereas others only make implicit reference to it, for example, in relation to electronics, control systems or programming (e.g. Australian Curriculum—Technologies, 2017; NAE, 2018; Technology in the New Zealand Curriculum, 2017).

In the light of the prominence of feedback in technology and engineering, it is surprising to note that there is virtually no research on how students of any age conceive of and/or learn about feedback in the technology and engineering education literature. There is an expanding literature on conceptions and learning of feedback in social and natural systems (e.g. Booth Sweeney & Sterman, 2000, 2007; Gilissen et al., 2019; Hmelo-Silver & Azevedo, 2006). There are also studies in undergraduate engineering education where students are subjected to, e.g. learning models based on dynamic engineering systems (Eftekhar & Strong, 1999), or systems thinking interventions in engineering team building (Walters et al., 2017), but very few deal with the learning of feedback in technological systems per se. One study, however, measured the perceived importance of introducing a haptic joystick for undergraduate students in a course on dynamic systems, and the students scored the highest on the use of the joystick in the laboratories about feedback control and virtual dynamic systems (Okamura et al., 2002). A few studies deal with feedback more generally in school and technology teacher education and report on both children and adult students' difficulties in understanding flows of information and feedback loops in technological systems (Hallström & Klasander, 2017; Koski & de Vries, 2013). Mioduser et al. (1996) conclude that in their study on middle school students' perceptions of control systems, the students "phrased the control laws in phenomenal or behavioral terms (what is observed), rather than in functional terms (how the system actually processes the information and produces outputs on acting components)" (p. 387). One of the underlying difficulties of understanding control features thus appears to be the fact that feedback loops are "invisible" in the sense that they are part of the flow of information in a technology or system, but more research is required about this important yet under-researched area, not only on the students themselves but also their teachers.

The aim of this chapter is to report a two-cycle intervention to improve Swedish secondary pre-service technology student teachers' conceptions of feedback in technological systems and to generalize some possible suggestions based on this study.

8.2 Feedback Understanding

The current study focuses on feedback as a curriculum component in secondary school and, more specifically, the teachers being trained to teach this content. This is why the intervention designed for the study—described in detail below—focuses on the student teachers' understanding of feedback. An analytical framework from a previous study about student teachers' conceptions of technological systems on a more general level (Hallström & Klasander, 2017) was employed to specify the systems-related nature of the students' answers. Hallström and Klasander (2017) identify three distinct, qualitative categories of systems understanding: atomistic, systemic and holistic conceptions. For this study, only the first and second categories, *atomistic* and *systemic*, were considered in the analysis because the holistic conception entails a comparison with other systems and the environment, which did not occur in the students' responses. Atomistic was made up of answers that reveal student conceptions of the parts or components in themselves, or that if one presses a button, then something happens, without making the systemic connection or seeing flows of information, matter or energy in the system. Systemic, on the other hand, included a conception of flows and connections that make components into a system, as well as a conception of the physical extension of the system (with wires, cables, pipes, etc.). Furthermore, this category included how the components interact more precisely and how flows of energy, matter and particularly information contribute to the working of the system, for example, through feedback loops (Hallström & Klasander, 2017, pp. 392–393).

As a way of describing the understanding of feedback in relation to the systemic and atomistic conceptions, one can say that systemic conceptions concern the basic, overall principles and features of feedback mechanisms such as the nature of negative and positive feedback and what role they play in different systems as well as how disturbances affect input, output and the stability of the system. The typical representation of the systemic level is the general block diagrams of the flow of information or energy in a technological system with feedback loop(s) (cf. Su et al., 2006). The atomistic conceptions, on the other hand, concern the component parts of the control systems with feedback loops, that is, what the central components are, their purpose and how they are connected (control unit/regulator, sensors, etc., represented with more detailed block/system diagrams). A good understanding of feedback mechanisms would require both systemic and atomistic conceptions, in line with classic cybernetics and control theory (cf. Hughes, 2004; Thomas, 2016; Wiener, 1948,

1950/1989). However, what the nature of these two types of conceptions is in the context of technology teacher education is an outcome of this chapter.

8.3 Methodology

The design of this study is a single arm, pre-post study design, a form of intervention study where one group is measured before and after an intervention. Since the study was carried out in two different cycles, it consequently includes two different groups but rather than the synchronous comparison common in, e.g. non-randomized trials, in this study the comparison is diachronous (Thiese, 2014).

8.3.1 Participants

The study was conducted at a Swedish university with student teachers studying to get a teacher degree in technology education and one or two other subjects: educational sloyd, mathematics education and/or science education (biology, chemistry and/or physics education). In the first cycle, four students studied for a teacher degree for lower secondary education (grades 7–9) and one student for upper secondary education (grades 10–12). This cycle included a pre-test carried out in the first session of an undergraduate course about teaching and learning about technological systems. A minor electable control theory component (excluding feedback) had been part of an earlier course, but in principle the pre-test was designed to capture the students' conceptions of technological systems and feedback before they had had any teaching about this, and before the intervention which was included as part of the current course. The post-test of the first cycle was carried out with four of the five students after finishing the course and thus after having gone through the intervention (one of the students was ill when the test was taken).

The second cycle was carried out with a new group of student teachers, in the same undergraduate course about technological systems but one year later. Then there were eight people studying to obtain a teacher degree for lower secondary education (grades 7–9), but two of them declined to take part in the second cycle of the intervention and the pre- and post-tests. Another student took part in the pre-test but did not follow through in the intervention and the post-test. Thus, there were altogether five students who completed the whole intervention including the pre- and post-tests. Table 8.1 summarizes the number of participating students in the two cycles of the intervention on feedback in technological systems. In both cycles, the two students participating in the pre-test but not the post-test are included in the analysis since it is the collective conceptions of the students before and after the intervention that are of interest.

Table 8.1 Number of participating students, female (F) and male (M), in the two cycles of the intervention on feedback in technological systems

Iteration\activity	Pre-test (F/M)	Intervention (F/M)	Post-test (F/M)
Cycle 1	5 (2/3)	5 (2/3)	4 (2/2)
Cycle 2	6 (1/5)	5 (1/4)	5 (1/4)
Total	11 (3/8)	10 (3/7)	9 (3/6)

8.3.2 Intervention and Data Collection

The undergraduate course on teaching and learning about technological systems was designed for pre-service secondary technology teachers and had five learning goals regarding systems theory and how it relates to a broad array of technological systems. Specifically, the intervention concerned the following learning goals: After finishing the course, the student should be able to 1. describe and explain functions of and relationships between technological components and systems of varying complexity, and 2. show in practical projects how control theory works in a defined system and relate this to teaching in secondary education (from course syllabus). The intervention in the first cycle consisted of a two-hour seminar, including a mini lecture by the researcher, about feedback in technological systems. The mini lecture and seminar were based upon a short text about feedback written by the researcher, "Feedback in technological systems" (see Appendix for a slightly revised English translation of the Swedish original). Furthermore, the intervention included a conference paper on the utilization of system models/block diagrams in learning about technological systems (Hallström et al., 2015) and a short, quite standardized textbook section about control systems and feedback for technology teachers in secondary education (Grimvall, 2014, pp. 127–133). The students were told to read the short researcher text as well as the paper and textbook section before the seminar so as to learn the appropriate concepts and to be prepared to discuss and ask questions about feedback at the seminar.

The second-cycle intervention was slightly modified to address issues that came out of the analysis of the tests in the first cycle. The second-cycle intervention was thus carried out with exactly the same mini lecture and seminar, short researcher text and literature apart from the fact that in the second cycle, Hallström et al. (2015) were discarded in favour of another short textbook section used in university-level courses about control theory and control systems (Thomas, 2016, p. 2–12). It addresses fundamental control theory and concepts, the design of block diagrams as well as the fundamentals of feedback, which were also focused on more in the mini lecture in the second round (see Fig. 8.1). In addition, in this second cycle, the students were tasked to design their own systems by creating their own block diagrams, in order for them to better understand the structure and design of feedback mechanisms.

Data were collected before (pre-test) and after (post-test) the interventions in both cycle 1 and cycle 2, through an anonymous questionnaire with two open-ended questions: "According to you, what is a technological system?" and "According

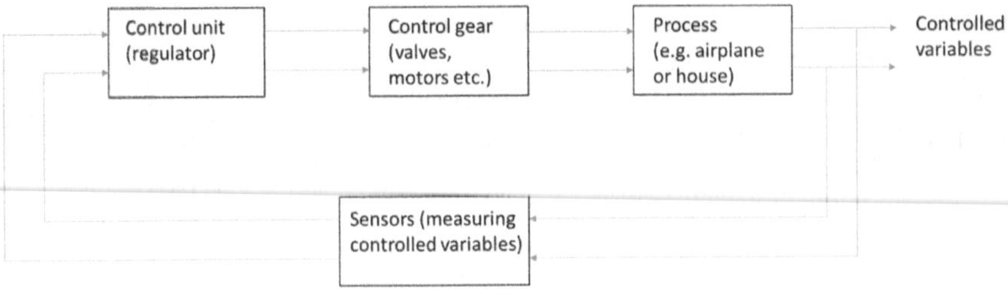

Fig. 8.1 Block diagram of control system with feedback, used in second-cycle intervention. *Source* Adapted from Thomas (2016), p. 5

to you, what is *feedback* in a technological system?". Only answers to the second question were analysed as part of this chapter. The questionnaire for the pre- and post-test in cycle 1 as well as the pre-test in cycle 2 was of a pen-and-paper variant so as to make it possible for the respondents also to draw, for example, a block diagram, in conjunction with their written responses. The questionnaire for the post-test in the second cycle had to be made digital and online as part of quarantine measures of Swedish universities in the spring of 2020 due to the coronavirus (COVID-19). This questionnaire was thus made in Google forms and distributed digitally to the respondents; it did not affect the response rate but restricted the responses to textual comments.

8.3.3 Data Analysis and Research Ethics

The collected data from the two rounds of pre- and post-tests were transcribed and subsequently analysed, coded and categorized using a hermeneutic method of text interpretation. This included repeated reading of the entire data set and generating codes in relation to relevant text sections, where single data snippets are continuously related to the whole data set of text and images in a re-interpretive way (Axell, 2015; Robson & McCartan, 2016). The analysis included a deductive phase in which conceptions about feedback that could be construed as *atomistic* and *systemic* were highlighted and further explored (Hallström & Klasander, 2017). The student responses were thereafter categorized so as to account for the qualitative elaboration of the conceptions of feedback, in relation to both atomistic and systemic conceptions. Categories were thus constructed inductively so as to account for the elaboration of the conceptions, and they came to be labelled *limited*, *intermediate* and *expanded* in relation to the potential for understanding feedback.

The rationale for the new, inductively created themes was that a respondent shows limited comprehension of feedback in a control system if she/he expresses 1. a very rudimentary conception of a feedback loop, and 2. that information about the actual result (output) is fed back into the system automatically to control it. A respondent shows intermediate comprehension of feedback in a control system if, in addition to what is listed above for limited comprehension, she/he expresses 3. that the purpose of employing feedback in automatic control (by using components such as sensors) is to create robust systems that react swiftly to disturbances, and 4. that negative feedback reduces fluctuations in the system's performance. A respondent shows expanded comprehension of feedback in a control system if, in addition to what is listed above for limited and intermediate comprehension, she/he expresses 5. that the system is controlled by way of a control unit or regulator that, based on the information from sensors, corrects the set point in order to get the desirable output/process value and 6. that when there is a disturbance, sensors feed back information to the control unit which then automatically adjusts the information accordingly so that the system performs its function with less and less fluctuation (negative feedback). With increasing elaboration of the responses from limited to intermediate to expanded, there is thus also a corresponding increase in the degree of complexity. However, a fourth theme, *undefined*, meant that any hint of a conception of a system, or a feedback loop, was lacking, or deemed unintelligible.

Throughout the research process, the ethical principles for research were followed. The student teachers consented to participation, after having been duly informed about the investigation in line with ethical guidelines (Swedish Research Council, 2017).

8.4 Findings

8.4.1 Pre-Test Cycle 1

As could be expected, when the students did the pre-test before starting the intervention, the overall understanding of feedback in technological systems was poor. One example of a limited conception was one which included the central control theory concepts of set point and process value but not much else:

> Feedback in a technological system is when the process value is compared to the set point, and the system is then adjusted to this so that the process value comes closer to the set point.

This conception was considered limited because, although the right concepts were used and it was systemic (a system is hinted to), neither the system, its components, nor how the system could adjust the process value by way of information from sensors, were clearly described. Another example was similarly labelled as limited:

> Feedback in a technological system may be when the various components in the system communicate with each other, "back and forth", in a kind of feedback, to improve the function of the system.

The above quote expresses a more atomistic conception, although a system consisting of components that communicate and feed back information in order to control it is alluded to. However, a feedback loop and important feedback concepts are missing, as are, for example, any mention of sensors (e.g. Goovaerts et al., 2019; Thomas, 2016). Below is one example of an undefined response:

> In the same way that control technology uses control to send back information from the system, feedback is needed in a technological system in order to make necessary changes in the system.

What this quote basically says is that feedback is the same as feedback. Also, while feedback indeed causes changes in the system, the direction of the feedback and the changes it causes are not specified.

8.4.2 *Post-Test Cycle 1*

There was a slight improvement of student understanding of feedback from the pre-test to the post-test, in that of all the responses, none was deemed undefined and even the intermediate conception could be found. An example of a post-test limited response, bordering on intermediate, is the following; it constitutes a rudimentary conception of a feedback loop and vaguely specifies a component/sensor:

> In a feedback loop information is given to earlier steps in the system. For instance, in district heating there are sub-stations that measure pressure and other kinds of information which then regulate the system so that the output is being kept invariant.

Another example of a more systemic limited conception is when a student defined feedback as "a way of controlling the system. Feedback can be found in different parts of systems". The student also drew a block diagram of a feedback loop, roughly the system model that was introduced in the intervention although it had more detail and complexity regarding how negative feedback is achieved (see Appendix and Grimvall, 2014). Yet another limited student response exemplified feedback with a water closet and how after flushing the tank is filled automatically, but the student failed to explain how the float acts as a sensor (see Fig. 8.2).

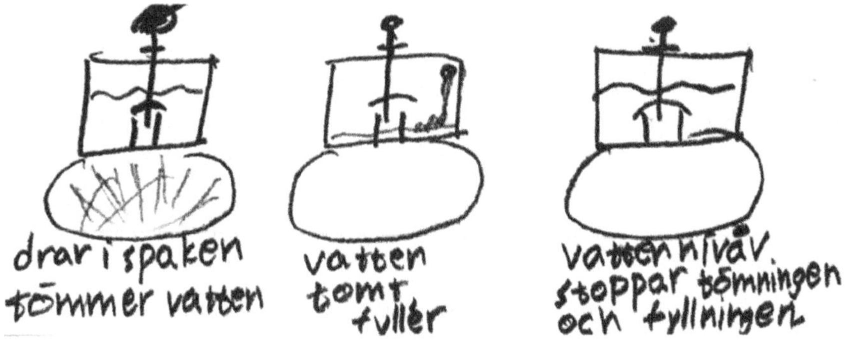

Fig. 8.2 Student-generated image of the flushing of a water closet

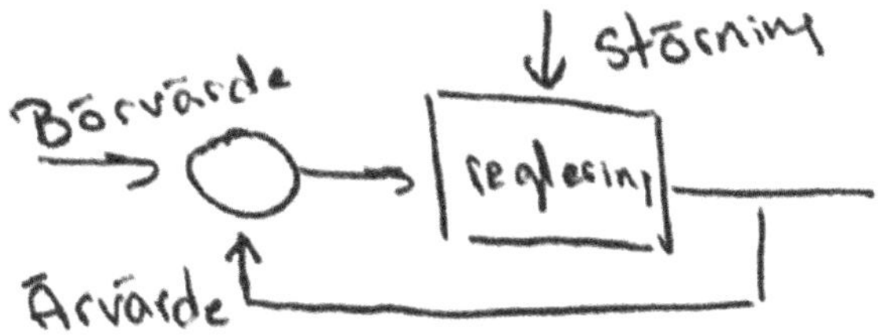

Fig. 8.3 Student-generated block diagram of a simple feedback loop

The water closet is also a closed system in the sense that unless there is leakage, there can be no disturbance. The feedback is therefore more sequential than controlling, as the float always lets in water when the tank is empty and stops when it is full.

An intermediate conception correctly represented the systemic perspective in the form of the block diagram of how a feedback loop works (see Fig. 8.3 and Appendix). This student also wrote:

> Positive/negative feedback, when "disturbance" happens in the system, components feed back in order to "stabilize" after "disturbance". For example, a change in the flow of matter, energy or information may require feedback for the flow to be able to continue circling in the system.

Thus, this student also expressed an atomistic conception detailing components/sensors, although it was rather vague.

8.4.3 *Summary of Findings Cycle 1*

The pre-service teachers taking part in this study had little or no prior knowledge of technological systems, control theory or feedback loops, so the fact that the group as a whole performed slightly better in the post-test than in the pre-test is not surprising. Although this is a small sample, the findings indicate that some student teachers understood the systemic aspects of feedback mechanisms, especially after the intervention. The systemic level of understanding technological systems, control and feedback has to do with the basic principles of feedback, as described in the basic block diagram that was part of the intervention (see Appendix). On the one hand, this could indicate that the intervention was successful for some students, especially since the basic principles of feedback are considered difficult to understand (Goovaerts et al., 2019; Martin, 1990). On the other hand, no student reached expanded comprehension and most conceptions were rather vague. Furthermore, there was a general lack of atomistic conceptions, relating to a micro-understanding. For example, although several students could at least vaguely express the basic principles of feedback, few could pinpoint sensors, how they work and their role in a control system. This study thus confirms findings of Mioduser et al. (1996) and Hallström and Klasander (2017) about the lack of essential *device knowledge* among both school students and adult student teachers.

8.4.4 Pre-Test Cycle 2

There were six new students doing the second cycle of the study, and just as in cycle 1, these students were essentially novices concerning the scientific areas of control theory and technological systems. The results of the pre-test were therefore, unsurprisingly, equally scarce compared to cycle 1: the student responses were judged as either undefined or limited. One student response, categorized as undefined, described what feedback in technological systems might be as "I actually have no idea. We haven't dealt with it in relation to technological systems with this term [feedback]. Context is what I remember. I do recognize the term though". Another undefined reply went: "To be influenced by, or to take part in, the technological system. To be able to use the technological system.". This is far too vague to give any indication of feedback, or what feedback is.

There was a variety among the limited responses, both as regard the nature of the descriptions of feedback and the orientation towards the systemic or atomistic aspects of the systems. The following response was deemed as limited in that it expressed a rudimentary conception of a feedback loop, although it was quite vague, bordering on the undefined:

> When a component can communicate to other parts of the system that can execute some kind of control, positive and negative feedback.

It was also atomistic in the sense that it focused on the components and parts, rather than on the system level and the principles of feedback. The text to the following limited response was scarce: "When a part of the system receives data based on its previous performance.". The student complemented it with a drawing (see Fig. 8.4).

The concepts used here were largely wrong: "Component →work→ Result → Analysis → feedback → Component". However, a rudimentary systemic conception of a feedback mechanism can still be detected.

Another student expressed a limited conception of feedback, at both the atomistic and systemic level:

> The input into a system can be used to correct the output, for example, a thermistor in a freezer which decides when the compressor cycle should start.

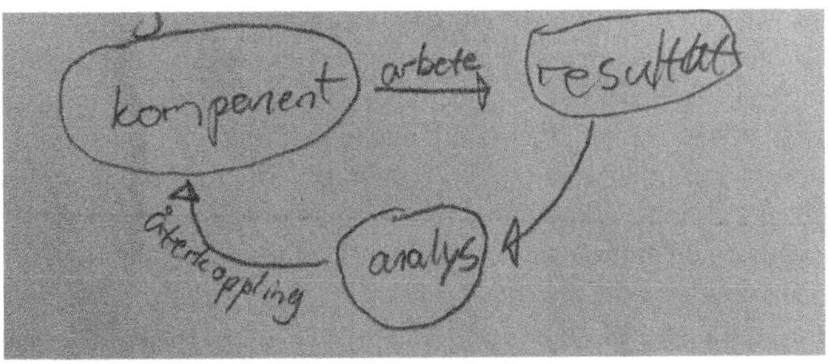

Fig. 8.4 Student-generated image of a conception of feedback

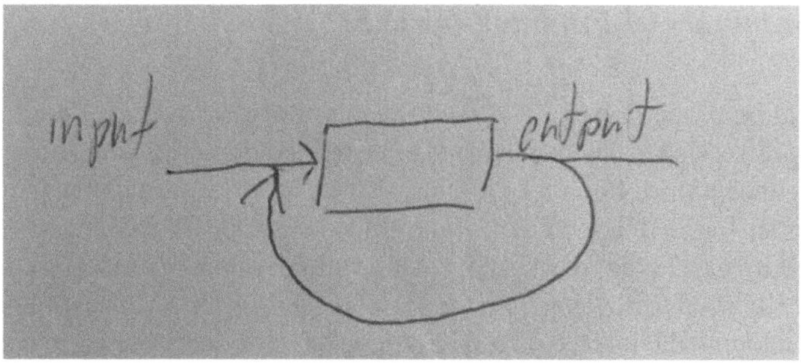

Fig. 8.5 Student-generated image of a simple feedback loop

This is really a misconception since in a control system, it is not the input that corrects the output, but the other way around. It is a sensor that feeds back information about a changed output (due to a disturbance, in this case a change of outer temperature) to a regulator or control unit which adjusts the freezer temperature accordingly. However, the student shows when mentioning the sensor (thermistor) and in the drawing a rudimentary/limited conception of feedback (see Fig. 8.5).

8.4.5 Post-Test Cycle 2

In contrast to cycle 1, there was little improvement of student conceptions of feedback from the pre-test to the post-test in cycle 2. There were, for instance, still undefined responses, one of which is the following:

> When the system monitors the input, for example, and takes measures.

The limited conceptions were all at a systemic level, that is, they concerned more the principles of feedback than the actual structure of the system and its components. Inclusion of more detail about the components of feedback loops in control systems could have made the following responses better reflect the workings of feedback:

> Feedback means that something is gauged/assessed, and this information is used to control or change something in the system.

> Some kind of information or similar in order to control, change or steer some part(s) of the system toward a desired goal.

An intermediate response took as its starting point a system model, input–process–output, and thus expressed a systemic conception that was considered more elaborate than the others in the post-test:

> If you see a system as input-process-output, feedback is evident because the output influences the input.

However, this conception lacks the atomistic level, and it is difficult so see from this response how a feedback loop works.

8.4.6 Summary of Findings Cycle 2

An implication of the first cycle of the research project was that for the subsequent cycle, the pre-service teachers may need to be trained more in device knowledge and atomistic conceptions of control systems, that is, how the various component parts work together for feedback to work. The students could, for example, be tasked to design their own technological systems with control features. Furthermore, since the students really need to develop both the systemic and atomistic conceptions, another way of developing the intervention might be to have students use more detailed and elaborated block diagrams (Thomas, 2016). Both a design task and the drawing of block diagrams were thus incorporated into the intervention in cycle 2.

Given the fact that the intervention had been improved in these respects compared to cycle 1, it is surprising that there was little improvement of student conceptions of feedback from pre- to post-test in cycle 2. The intervention in cycle 2 was more grounded in tertiary-level control theory literature, and the students were given the opportunity to learn how to draw block diagrams of simple control systems. One reason for the little improvement between pre- and post-test in cycle 2 might be the constricted form of the post-test in Google forms, since the students could not elaborate on details of the feedback mechanisms by drawing block diagrams. It is also a distinctive feature of the responses in cycle 2 that they are mostly systemic and not very atomistic.

8.5 Concluding Discussion and Implications

At first glance, it may be considered discouraging that there was such a slight improvement of secondary technology student teachers' conceptions of feedback in technological systems especially after the second cycle, despite prior efforts to improve the intervention. However, in both cycles, there was some improvement. The participating students were also complete novices, and the intervention just consisted of a two-hour seminar and some preparatory reading. Consequently, this could still be considered a good start towards improving students' conceptions of feedback in technological systems, and it is encouraging that many students could develop a systemic understanding. However, further studies need to be done to investigate how students' atomistic conceptions and device knowledge could be improved (Mioduser et al., 1996). Perhaps this could be done along the lines of the second-cycle intervention of this study, with focus on creating block diagrams and designing simple technological systems with control mechanisms, although it needs to be more strictly planned, monitored and evaluated.

There is also another way in which the findings of this study could be seen as encouraging. Previous studies of feedback show that students of all ages find it very difficult to understand nonlinear systems, in general, and feedback in particular (e.g. Arbesman, 2017; Hmelo-Silver & Azevedo, 2006). Thus, although the intervention and research design of the current study can be improved, there is also an inherent difficulty in working to improve students' conceptions of complex systems because they generally find the systems complicated and counterintuitive to comprehend. Indeed, many researchers agree that the most complex contemporary systems cannot be fully understood by a single person, or, sometimes, even by a combination of several experts (cf. Arbesman, 2017; Hughes, 2004). So, if an intervention like the present one can lead to some improvement in students' understanding of feedback in technological systems and their more general systems thinking skills, it is yet an important step forward.

The findings of this study could also serve as pointers for technology teachers and teacher educators on how to teach about feedback and control. First of all, even though the student teachers found the subject of feedback in technological systems difficult, if we look at the examples given by the students themselves in the pre- and post-tests, they are generally rather simple, well defined or well known. In a similar vein, Hallström and Klasander (2020) suggest that teachers choose examples of technological systems along the following progression lines, in order to promote the most efficient learning:

- From simple to complex systems,
- From small to large and widespread systems,
- From systems related to myself via us to others,
- From local systems via regional/national to global systems (Hallström & Klasander, 2020, p. 79).

The student teachers' choice of simple, small, local, etc., systems may be one reason that they actually improved their understanding of feedback somewhat. Secondly, although often limited, students' conceptions of systemic aspects of feedback seem to have been easier to come to grips with and improve than the atomistic ones (cf. device knowledge, Mioduser et al., 1996). This may indicate that it is a more efficient pedagogical strategy for teachers to focus on the overall principles of feedback before delving into different components, sensors and devices in a control system.

Further ways for teachers to develop students' understanding of feedback, especially when they have passed the limited and intermediate conceptions, may be to focus on the mathematics involved in feedback in simple, static control systems, e.g. proportional feedback, and, further, in dynamic systems. Apart from learning the mathematical calculations, students are also forced to reflect on both atomistic and systemic aspects of such systems and included feedback mechanisms. Discussion of complex control problems and rules of thumb may also ensue (Glad & Ljung, 2006; Norström, 2014). Such more advanced projects can also develop students' modelling skills and integrated STEM literacy, something which, in turn, is very important for learning more advanced control theory (Hallström & Schönborn, 2019; Tang & Williams, 2019).

8.5.1 *Limitations*

The cycle 2 post-test questionnaire made in Google forms precluded any drawing on the part of the students. This may have affected the findings from cycle 2 in that the students could not draw block diagrams and thereby show how components are coupled together in feedback loops, that is, there may have been a skewness in the answers towards the systemic conceptions that are perhaps more easily describable in words. Typically, atomistic conceptions can be shown by drawing a detailed block diagram. Furthermore, it is conceivable that the broad question "According to you, what is *feedback* in a technological system?" may have determined, at least to some extent, the overall degree of complexity of the student answers and thus also what category from limited to expanded they fell into. In future studies of feedback, it might be useful to include different kinds of questions in order to capture this elusive phenomenon. This study generated some findings, but it also points forward towards future, in-depth studies with improved research instruments.

Appendix

Feedback in Technological Systems

Most technological systems make use of various types of feedback to control the system in as effective a way as possible, but also to make the system automatic. Feedback loops make the system automatic, because it then regulates itself with the help of different kinds of sensors. Feedback exists in different kinds of systems—social, ecological, technological, etc.—which is why it is important to know how it works. The purpose of this text is to describe feedback in technological systems in a more general way than is common in the literature, so as to be appropriate for pre-service technology teachers in secondary education.

What is Feedback?

Feedback features in all kinds of systems. An everyday example is the sale of milk cartons in a grocery store. The shopkeeper orders milk cartons from a supplier, to be delivered once or often several times a week. When the demand is high, as, for instance, before Christmas, the shopkeeper orders more milk than usual, but when the demand is low and the corresponding milk sale also low, the shopkeeper needs to order less milk to balance the stock in the store. In this way, the inflow in the milk sale system is controlled by the milk orders from the grocery store, and the shopkeeper is a kind of "sensor" which monitors the stock of milk cartons. The system is controlled towards an optimal stock that matches the demand for milk, and so the feedback has

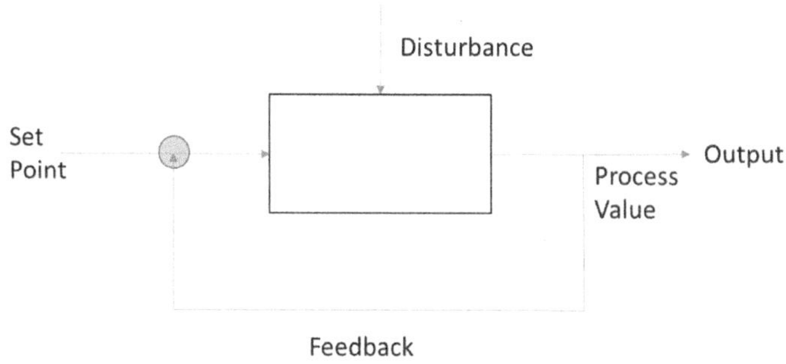

Fig. 8.6 Block diagram/system model of the flow of information in a feedback loop. *Source* Inspired by Grimvall (2014), p. 129

a dampening effect—it counteracts too big fluctuations in the milk orders, from the point of view of the grocery store. This is called *negative feedback* and is the most common type of feedback.

An example of *positive feedback* is when many people in a room are talking at the same time, and each person needs to raise their voice little by little, in order to be heard. This leads to an amplifying effect, the opposite of a dampening effect (cf. Levary, 1986). The most common form of feedback in technological systems, however, is negative feedback, so that is what we focus here.

Feedback in a technological system is thus a way of controlling it with sensors that feed back information to the system's control unit (Mioduser et al., 1996). Figure 8.6 shows a way of describing the *flow of information* in a feedback loop in a technological system.

What Does Feedback Look like in Different Technological Systems?

When describing in more detail how feedback mechanisms are designed in technological systems, one cannot rely only on the above block diagram (Fig. 8.6) because it merely shows the flow of information and how it contributes to the control of the system. A technological system with feedback loops can be designed in various ways and has many different types of flows, both of information (e.g. the Internet), matter (e.g. water supply) and energy (e.g. electric grids). Below is a system diagram of a simple control system with feedback (Fig. 8.7) in which the focus is on how the system is physically connected to achieve control.

Several different systems that look like Fig. 8.7 are conceivable. For example, we have a motor whose speed is fed back to the control unit, with possible manual control using a potentiometer or automatic control employing a computer programme. Disturbances, for instance, something hindering the rotation of the motor, are fed back through the sensor and can be adjusted manually or automatically by increasing

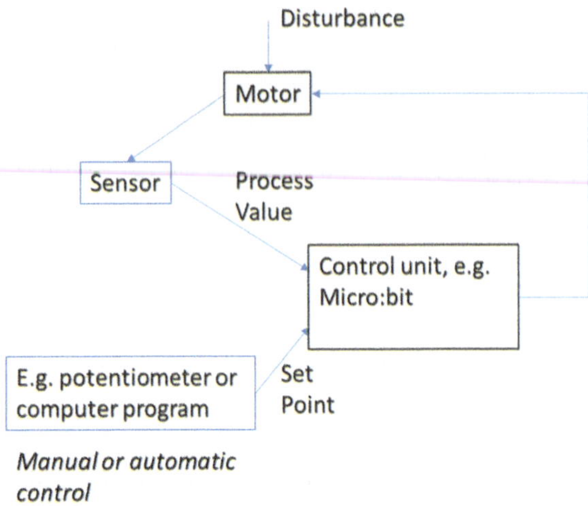

Fig. 8.7 System diagram of a simple control system with manual or automatic feedback

the speed via the control unit. It is also conceivable that the block diagram shows the cruise control in a car, in which the control is automatic and continuous. The driver adjusts the desired speed manually, but software in the car makes sure that the speed is even using a negative feedback mechanism. Disturbances happen continuously as the strain on the engine changes due to the varying topography of the landscape on which road is built. The software makes sure that the engine works harder/less hard and that the speed thus increases/decreases depending on whether the speed is lower or higher than the set point.

References

Arbesman, S. (2017). *Overcomplicated: Technology at the limits of comprehension*. Portfolio.

Australian Curriculum—Technologies. (2017). Retrieved Dec 6, 2019 from https://www.australiancurriculum.edu.au/.

Axell, C. (2015). *Barnlitteraturens tekniklandskap: en didaktisk vandring från Nils Holgersson till Pettson och Findus (Technology landscapes in children's literature: A didactic journey from Nils Holgersson to Pettson and Findus)*. Dissertation. Linköping: Linköping University.

Barak, M. (2018). Teaching electronics: From building circuits to systems thinking and programming. In M. J. De Vries (Ed.), *Handbook of Technology Education* (pp. 337–360). Springer.

Booth Sweeney, L., & Sterman, J. D. (2000). Bathtub dynamics: Initial results of a systems thinking inventory. *System Dynamics Review, 16*(4), 249–286.

Booth Sweeney, L., & Sterman, J. D. (2007). Thinking about systems: Student and teacher conceptions of natural and social systems. *System Dynamics Review, 23*(2/3), 285–312.

NAE. (2018). *Curriculum for the compulsory school, preschool class and school-age educare, revised 2018*. National Agency for Education.

Eftekhar, N., & Strong, D. R. (1999). Towards dynamic modeling of a teaching/learning process Part 3: The simulation model. *International Journal of Engineering Education, 15*(3), 168–190.

Gilissen, M. G., Knippels, M. C. P., Verhoeff, R. P., & van Joolingen, W. R. (2019). Teachers' and educators' perspectives on systems thinking and its implementation in Dutch biology education. *Journal of Biological Education* 1–12, https://doi.org/10.1080/00219266.2019.1609564.

Glad, T., & Ljung, L. (2000). *Control theory: Multivariable and nonlinear methods*. Taylor & Francis.

Glad, T., & Ljung, L. (2006). *Reglerteknik. Grundläggande teori*. Studentlitteratur.

Goovaerts, L., De Cock, M., Struyven, K., & Dehaene, W. (2019). A concrete proposal to introduce control theory to 16 year old pupils. *European Journal of STEM Education, 4*(1), 1–11.

Grimvall, G. (2014). *Teknikens metoder. Skolans teknikämne i senare skolår*. Studentlitteratur.

Hacker, M. (2018). Engineering and technology concepts: Key ideas that students should understand. In M. J. De Vries (Ed.), *Handbook of technology education* (pp. 173–192). Springer.

Hallström, J., & Klasander, C. (2017). Visible parts, invisible whole: Swedish technology student teachers' conceptions about technological systems. *International Journal of Technology and Design Education, 27*(3), 387–405.

Hallström, J., & Schönborn, K. J. (2019). Models and modelling for authentic STEM education: Reinforcing the argument. *International Journal of STEM Education, 6*, 22.

Hallström, J., & Klasander, C. (2020). Making the invisible visible: Pedagogies related to teaching and learning about technological systems. In P. J. Williams & D. Barlex (Eds.), *Pedagogy for Technology Education in Secondary Schools: Research Informed Perspectives for Classroom Teachers* (pp. 65–82). Springer.

Hallström, J., Klasander, C., & Svensson, M. (2015). The black box and beyond: Introducing a conceptual model as a learning tool for developing knowledge about technological systems. In M. Chatoney (Ed.), *PATT 29 Plurality and Complementarity of Approaches in Design & Technology Education, Marseille, France, April 2015*. Presses Universitaires de Provence.

Hmelo-Silver, C. E., & Azevedo, R. (2006). Understanding complex systems: Some core challenges. *The Journal of the Learning Sciences, 15*(1), 53–61.

Hughes, T. P. (2004). *Human-built world: How to think about technology and culture*. University of Chicago Press.

International Technology Education Association. (2007). *Standards for Technological Literacy: Content for the Study of Technology* (3rd ed.). International Technology Education Association.

Koski, M.-I., & de Vries, M. J. (2013). An exploratory study on how primary pupils approach systems. *International Journal of Technology and Design Education, 23*(4), 835–848.

Levary, R. R. (1986). Simulation as a tool for the analysis and design of systems. *European Journal of Engineering Education, 11*(2), 165–176.

Martin, D. J. (1990). Stages in systems thinking, modelling and realisation in school electronics. *International Journal of Technology and Design Education, 1*(1), 14–20.

Mindell, D. A. (2002). *Between human and machine: Feedback, control, and computing before cybernetics*. The Johns Hopkins University Press.

Mioduser, D., Venezky, R. L., & Gong, B. (1996). Students' perceptions and designs of simple control systems. *Computers in Human Behavior, 12*(3), 363–388.

Norström, P. (2014). *Technological Knowledge and Technology Education* (diss). Royal Institute of Technology (KTH).

Okamura, A. M., Richard, C., & Cutkosky, M. R. (2002). Feeling is believing: Using a force-feedback joystick to teach dynamic systems. *Journal of Engineering Education, 91*(3), 345–349.

Robson, C., & McCartan, K. (2016). *Real world research: A resource for users of social research methods in applied settings* (4th ed.). Wiley.

Su, J.-H., Chien, C.-L., Chen, J.-J., & Wang, C.-M. (2006). Simulink behavior models for DC-DC switching converter circuits using PWM control ICs. *International Journal of Engineering Education, 22*(2), 315–322.

Swedish Research Council. (2017). *Good Research Practice*. Stockholm.

Tang, K.-S., & Williams, P. J. (2019). STEM literacy or literacies? Examining the empirical basis of these constructs. *Review of Education, 7*(3), 675–697.

Technology in the New Zealand Curriculum. (2017). Retrieved Dec 6, 2019 from http://nzcurriculum.tki.org.nz/.

Thiese, M. S. (2014). Observational and interventional study design types; an overview. *Biochemia Medica, 24*(2), 199–210.

Thomas, B. (2016). *Modern reglerteknik*. Liber.

Walters, J. P., Greiner, B., O'Morrow, E., & Amadei, B. (2017). Fostering systems thinking within engineers without borders student teams using group model building. *International Journal of Engineering Education, 33*(1), 247–260.

Wiener, N. (1948). *Cybernetics: Or control and communication in the animal and the machine*. Hermann & Co., Editeurs.

Wiener, N. (1950/1989). *The human use of human beings: Cybernetics and society*. Free Association Books.

Jonas Hallström is Ph.D. and Professor of Technology Education at the Department of Behavioural Sciences and Learning, Linköping University, Sweden, where he is also currently Deputy Head of Department. He is also Chair of the Swedish National Graduate School in Science and Technology Education Research (FontD). He presents regularly at international conferences and consults on technology education. He is one of the editorial board members of the Brill/Sense book series *International Technology Education Studies*. His research primarily concerns the historical emergence of technology as knowledge content in the school, the epistemology and subject philosophy of technology, various subject content (e.g. technological systems) as well as the attitudes to and knowledge of technology and technology education of students, student teachers and teachers. The research also relates to technology teaching in relation to, e.g. design, gender (girls and technology), authentic learning, models and modelling and STEM (science, technology, engineering, mathematics) education.

Chapter 9
Student Teachers' Mental Models of Everyday Adaptive Control Systems

Osnat Dagan

Abstract Use of adaptive control systems is constantly increasing, and more technological systems are becoming self-regulated. The understanding of how they work becomes more difficult as the processes are invisible, like a black box. The aim of this chapter is to understand what people's prior mental models of everyday adaptive control systems are (from the literature and from local research) in order to develop an instructional unit that helps student teachers construct their mental models of control systems. Undergraduate kindergarten and primary school student teacher's at Beit Berl College (BBC), Israel attend the "Technological thinking and robotics" course in which they learn about everyday adaptive control systems. It became clear to us that after the course students still do not have correct mental models of such systems that could assist them in life and as teachers. Therefore, the prior mental models of student teachers were examined. They were asked to describe how an automatic door works, either in sketches or in writing. Based on the findings arising from the data, which reinforce earlier research findings (Hallström and Klasander in International Journal of Technology and Design Education 27:387–405, 2017; Koski and Vries in International Journal of Technology and Design Education 23:8, 2013; Slangen et al. in Professional development for primary teachers in science and technology. 2011; Svensson and Klasander Technology Education Research Conference, Surfers Paradise, Australia, 2012), an instructional unit containing four activities has been developed. These activities' main aim is to make the invisible visible, in order to clarify how self-regulation control systems work and help student teachers construct accurate "runnable" (Mioduser et al. Computers in Human Behavior 12:363–388, 1996) mental models.

Keywords Control mental models · Higher education · Adaptive control systems · Pre-service teacher training · Instructional unit

---

O. Dagan (✉)
Beit Berl College, Beit Berl, Israel
e-mail: osnat.dagan@beitberl.ac.il

J. Hallström and P. J. Williams (eds.), *Teaching and Learning about Technological Systems*, Contemporary Issues in Technology Education,
https://doi.org/10.1007/978-981-16-7719-9_9

9.1 Introduction

Today's world is filled with "smart" control systems that behave according to their surroundings, adapting their behaviour as their sensors indicate the changes around them, for example: automatic doors, air-conditioners, smart ovens, adaptive cruise control, etc. These control-smart (adaptive) systems can achieve a desired behaviour while their surroundings change rapidly.

Student teachers who will teach technology education including control systems and processes are expected to understand how these smart, adaptive, self-regulation systems work, possess the correct mental models and pedagogical content knowledge (PCK) and be able to assist their future pupils to construct their own mental models. Thus, it is important to analyse student teacher's prior mental models in order to design an appropriate instructional unit to equip them with the correct mental models.

There are only few studies on mental models of systems and control systems among K-12 pupils and student teachers. This chapter presents a small study that supports earlier ones focused mainly on children (Koski & De Vries, 2013; Lind et al., 2019; Mioduser et al. 1996) and less on student teachers and active teachers (Hallström, & Klasander, 2017; Koski & De Vries, 2013; Slangen, Van Keulen & Gravemeijer, 2011; Svensson, & Klasander, 2012). The Beit Berl College research (Dagan, 2021), reported in this chapter, found that student teacher's control process mental models for everyday adaptive systems is partial; most of them have a structural model that contains only the visible components belonging to the operation unit and not the control unit of the system. Their functional mental models are also lacking. However, most of them explain the behaviour of the system as a script of sequential process (a, b, c…) and with rules (If…then …). Their PCK is partial and insufficient for teaching.

An instructional unit was developed on the basis of the conclusions and insights from research studies on self-regulation control systems. This unit contains four activities, from observation to analysis and then to designing, building and programming, scaffolding students from concrete to the abstract thinking.

The aim of this chapter is to identify from prior studies and from the current research data what student teacher's prior adaptive control systems mental models are, and to design an instructional unit that will bridge the gap between these mental models and accurate ones.

9.2 Literature Review

The literature review below relates to the concepts of control systems, the mental models created by pupils and teachers, and the teacher's knowledge.

9.2.1 *Models of Self-Regulation Control Systems*

Teaching technological systems is complex. A technological system is defined as a collection of components that work together in coordination in order to fulfil the system's objectives. Each system has input, process and output. The input could be energy, material and information. The output impacts the environment and society. A system contains sub-systems and mechanisms that could be described in a block diagram. Systems thinking includes understanding of the following concepts: structure, factors that impact the output, the big picture "macro view", boundaries, function and behaviour, feedback, dynamics and safety (Barak, 2018, p. 342). Figure 9.1 is an example of a technological system block diagram taken from a technological systems textbook for 8th grade developed by ORT Israel (Kipperman et al., 2000).

The technological systems in our everyday lives tend to be taken for granted and have almost become invisible. Many of them are controlled by self-regulating sub-systems acting as feedback loops that are capable of modifying their own behaviour to accommodate an unexpected change (Rzevski, 1995). The system regulates its behaviour by comparing data from the environment (collected by sensors) with the desired output (a programmed algorithm that is part of the controller) and sending instructions to the operators (light, motors, etc.) to act in order to obtain the desired output.

Control systems are divided into two units: operation (OU) and control (CU) (Levin & Mioduser, 1996) (Fig. 9.2). The OU is further divided into two: the perception sub-system that collects data on the system and its environment state, and the

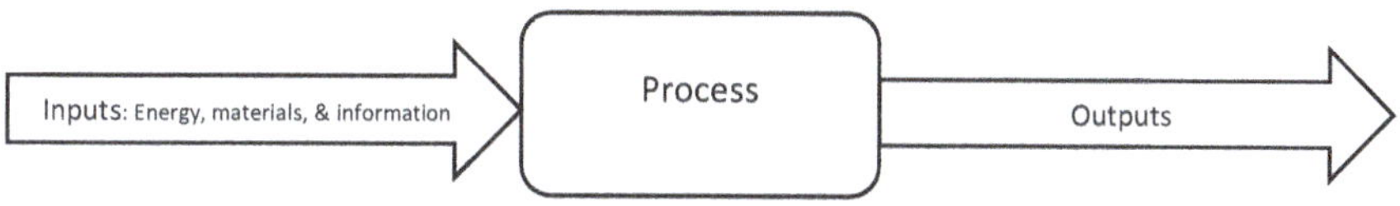

Fig. 9.1 A block diagram of technological system

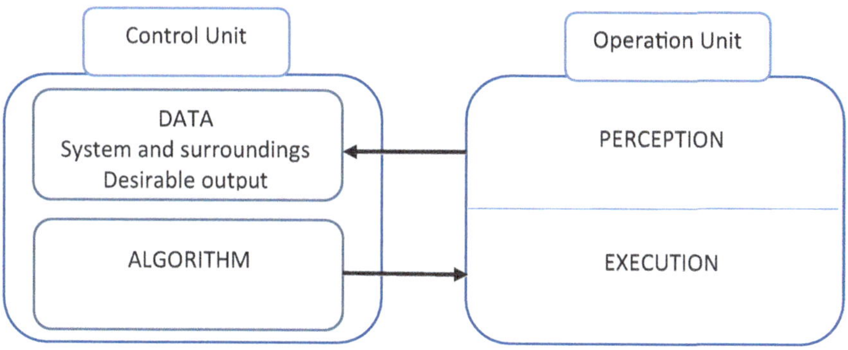

Fig. 9.2 Schematic chart of control system (based on Levin & Mioduser, 1996)

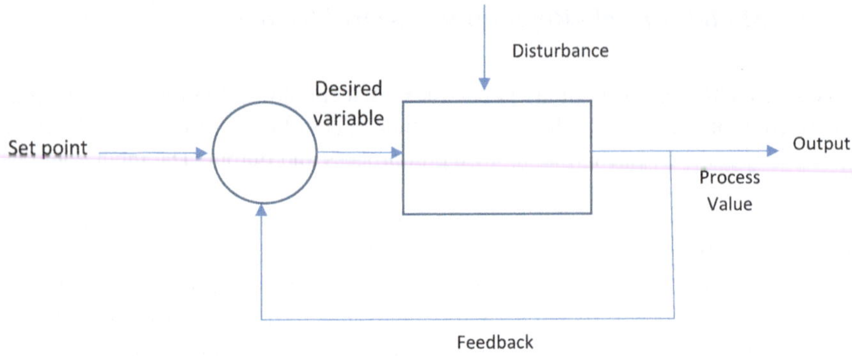

Fig. 9.3 Self-regulation process in adaptive system—the flow of information in the feedback loop (based on Hallström, 2021)

execution sub-system that starts and stops the system operators (lights, motors, engines etc.). The control process—information (value of measures) collected by the perception sub-system received from the OU—is compared to the desired output by operating an algorithm. A decision on the continuing action is made and sent to the execution sub-system in order to obtain the desired output. The CU's "aim" is to cause the desirable behaviour of the system as defined in the algorithm within the self-regulation of the system. The entire process can take place with no human involvement (Fig. 9.2).

Engineers describe the control system as a self-regulation system, a closed-loop system (Fig. 9.3 based on Hallström, 2021). This block diagram shows the information flow and the comparison between the set point and the process value, which engineers call a "closed loop".

From our prior experience, we have understood that the traditional way of presenting the closed-loop feedback in a block diagram (Fig. 9.3) is too complicated, not only for pupils, but also for primary and even some lower secondary teachers. A pedagogical model of control in adaptive, self-regulated systems describing the conceptual framework was developed from the traditional model (Fig. 9.3) integrated with the schematic chart (Fig. 9.2) in the ORT Israel textbook (Kiperman et al., 2000), as shown in Fig. 9.4. This model was chosen because it describes the control process in a detailed and qualitative manner.

Self-regulation in the CU (indicated in Fig. 9.4 as A, B and C) is an information process. The sensors from the OU collect information about the current output of the system and its surroundings and transfer it (by CU A) to B—the algorithm. B. gets two inputs, one is about the desirable output and the other one comes from the sensor (through A). The algorithm compares these two inputs according to the rule: If… then… / If not… then…and transfers the decision through C to the OU operators in order to correct deviations. For example, in an adaptive air-conditioning system, the desirable system output is that the temperature in the room will be 25 °C throughout the day. The sensor that is part of the OU measures the temperature in the room

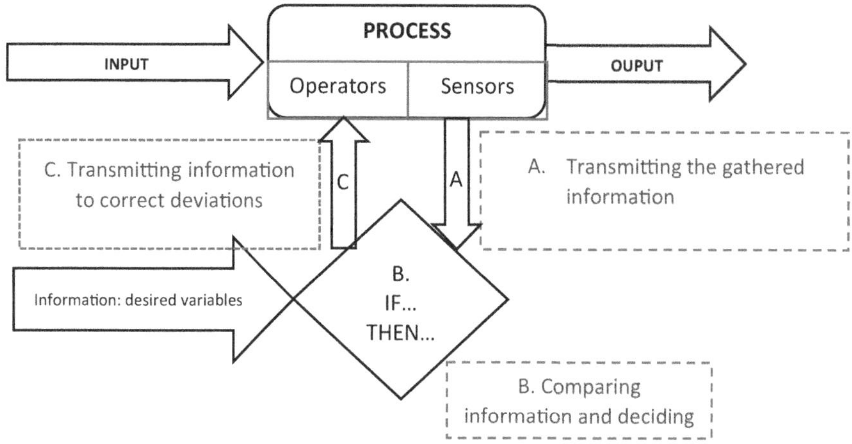

Fig. 9.4 Self-regulated systems model (Kiperman et al., 2000)

and transmits the measurements to the CU (through A). The controller compares the measurements with the desirable output with its algorithm (B). If the temperature is higher than 25 °C, it transmits information to the operator to cool the room (C). If the temperature is 25 °C or lower, then it transmits information to the operator to stop cooling. The temperature in the room will remain around 25 °C all the time, as shown in Fig. 9.5.

The first presented model (Fig. 9.2) is a conceptual model of an adaptive control system, the second (Fig. 9.3) is the model in use in engineering and in high school. The third (Fig. 9.4) was developed from both as a pedagogical model for younger pupils (K-9) in order to help them to develop the correct mental models of such systems.

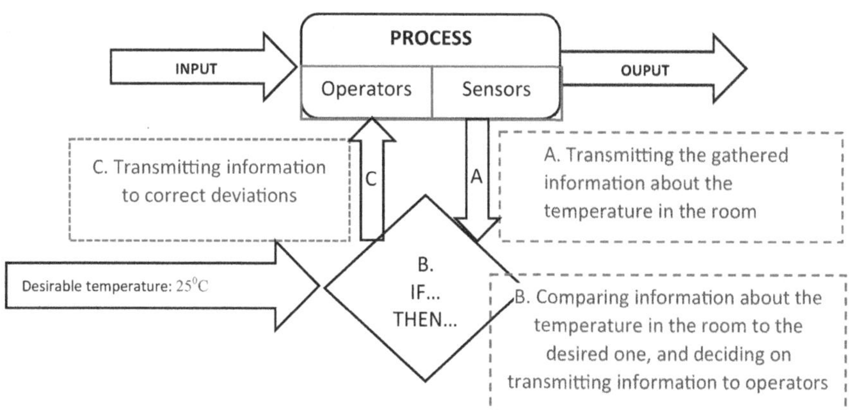

Fig. 9.5 An example of a self-regulated air-conditioning system

9.2.2 *Pupil's and Teacher's Mental Modelling of Systems and Control Systems*

It is considered important to learn about feedback as part of technology literacy and of understanding the world around us (Barak, 2018), but it is regarded as a difficult subject. It used to be introduced to students only in secondary school. Now that robotics and computational thinking are part of the compulsory K-6 curriculum, it must be taught much earlier.

Mental models are internal cognitive representations of real-world situations that assist description, explanation and prediction (Norman, 1983). Mental models, including system mental models, are constructed step-by-step during interactions between a person and a system and in layered fashion (every day, over years and multiple interactions). This model changes and develops according to the knowledge one possesses while experiencing the system. This layered mental model is constructed from concrete to abstract to causal (Norman, 1983). A mental model is used to understand, explain, troubleshoot and improve a system or design a new one (Mioduser et al., 1996; Norman, 1983). Usually, a system's mental model is intuitive, partial and inaccurate, and people are limited in their ability to "run" it in their head in a way that matches reality (Norman, 2014). As a result, and given that technological systems are complex and hidden, people's mental models tend to vary, often including misconceptions of how the different layers of the system work. Jones (2003) argued that the understanding of systems is essential in developing knowledge in technology (Jones, 2003, p. 90). By system mental models we mean systematic structural / functional / causal cognitive models (Mioduser, 1998; Mioduser & Dagan, 2007). Most research studies about mental modelling in technological systems relate to (a) people's understanding of the way these systems work; and (b) their use of, and problem solving with, these systems (DeKleer & Brown, 1981; Kieras, 1988; Norman, 1983; White & Fredriksen, 1986). The main findings in these studies indicate that students who possessed an appropriate structural and functional mental model of a given system used it to design and plan effective solutions to problems related to the system's functioning or operation (Kieras, 1988) and troubleshooting (White & Frederiksen, 1986). These mental models assisted people in producing causal explanations (White & Frederiksen, 1986) and in making predictions about novel situations not yet examined (Kieras & Bovair, 1984).

Research on technological system perceptions, and particularly control systems, is very limited for children and adults alike. Studies on systems understanding among primary school pupils and teachers, found that (a) the concept of input was clearer to pupils than output; (b) the pupils had a linear conception of systems; (c) pupils are better at understanding the structure of a system than its behaviour, control mechanisms and flow of information; (d) the challenge was to differentiate between a process and a system, the role of information in the system and setting boundaries for the system; (e) pupils also better understand systems when they are scaffolded; (f) teachers need more knowledge about the similarities and differences between various technological systems and better understanding of system components and different layers; and (g) teachers mainly focused on system structure, input, process, output and consequences for the environment rather than on the system behaviour, the control mechanisms and the flow of information (Hallström, 2021; Koski & de Vries, 2013; Lind et al., 2019; Schooner, Klasander, & Hallström, 2015; Slangen et al., 2011; Svensson & Klasander, 2012).

Findings from research on student teacher's perceptions of technology systems showed that (a) most student teachers could see the various parts in the system but were unable to connect them to a wider context; (b) almost half the students provided answers that were considered undefined; (c) the parts of the systems the students understood were mostly the visible ones (Hallström, & Klasander, 2017).

Mioduser et al. (1996) investigated mental models that American middle-school pupils produce of control systems such as automatic doors and household devices, before, during and after instruction. They studied pupils' "conceptions, missing conceptions and misconceptions" of these control systems. They developed a sequence of qualitative models of the control system consisting of four types of models, from a phenomenological and general model to a causal model: (a) black box model—describes the overall behaviour of the system, indicating that in the presence of input, it produced an output. The process, the structural and the functional aspects of control process are ignored; (b) reactive model—sensing functions and devices are explicitly mentioned. The system is identified as a "sensing acting device"; (c) Switch model—separate command-delivering functions appear, reflect the awareness of a needed control unit for decision making, but the nature of the controlling function remains undefined; and (d) control model—control specifications are included and even represented in some formal way of algorithm. Causal relationships in the system functioning appears (Fig. 9.6) (Mioduser et al., 1996 p.15–16).

Their conclusions were that the pupils' understanding prior to instruction was very poor (but slightly better after). They used mostly "black box" and "reactive" models and only rarely the "switch" and the "control" models. They lacked accurate understanding of common components and how they affect the system. System structure was well understood, while control features of the system were poorly understood, as was the flow of information in the system (Mioduser, et al., 1996).

Following on from this research, in recent years studies have been conducted at Tel-Aviv University on kindergarten and primary school children's mental models of self-regulated systems. These studies found that kindergarten children understood

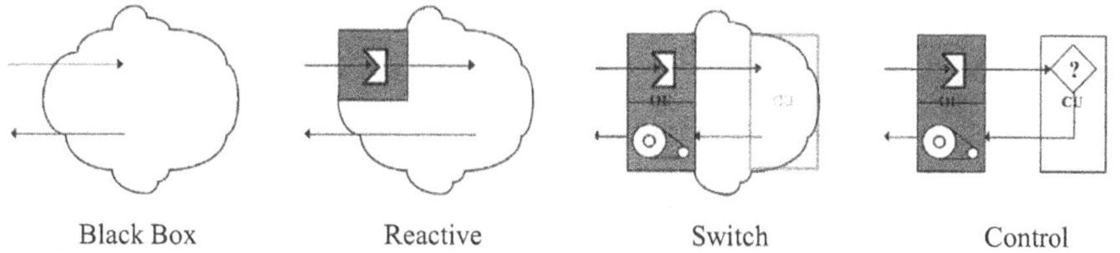

Fig. 9.6 Sequences of qualitative models of the control system (Mioduser et al., 1996 p.15)

the input unit, and that there are rules determining an automatic door's operation, but did not understand "who" really operates them. These misunderstandings led them to misconceptions and incorrect explanations (Rodan, 2016). The pupils' mental models do not include the CU, probably because it is invisible, as was found by Hallström and Klasander (2017) with adults. Moreover, K-6 pupils' prior perceptions of control systems were mostly "black box" and after experiencing and learning with adaptive robots were based on "switch" and "control", which are considered quite advanced (Landsman, 2019; Rodan, 2016). Furthermore, while working with young pupils it was found that (a) the scope and type of interaction have a crucial impact on the construction of the mental models of control systems; (b) acting with an adaptive robot enhances the generalization of the robot behaviour and (c) teacher's support and scaffolding is essential (Mioduser et al, 1996; Rodan, 2016).

Hallström (2021) researched student teacher's conceptions of feedback in technological systems based on three distinct, qualitative categories of system understanding: atomistic, systemic and holistic conceptions (Hallström & Klasander, 2017). This analysis leads to the understanding of the two ways of reaching conceptual comprehension about feedback in technological systems: macro and micro. A full understanding of feedback mechanisms requires both macro and micro understanding. The conclusions from his research are: (a) the intervention helped some of the student teachers understand the systematic (macro) aspects of feedback mechanisms; (b) none of the student teachers reached expanded understanding and most of their conceptions remained vague; (c) they all lacked atomistic conceptions (micro understanding).

9.2.3 Teacher's knowledge

For most of the population, technological devices and systems, and especially adaptive, self-regulation systems, remain a black box. These systems are all around us in our daily life. That is why people need some scientific and technological literacy, in order to not just push the button in order to operate the device, but to be aware of its underlying processes.

In order to help pupils become technologically literate, aware of the control processes and be able to construct the right control mental models, we need to be sure that their teachers possess the appropriate subject matter knowledge (SMK), pedagogical knowledge (PK) and the pedagogical content knowledge (PCK).

To be effective, teaching requires basic skills, content knowledge and general pedagogical skills and begins with teacher's understanding of what is to be learned and how it is to be taught (Shulman, 1987, p. 7). Based on Shulman's (1987) criteria of teacher's knowledge base, Grossman (1990) designed a model of teacher knowledge that includes: (1) SMK; (2) General PK—the understanding of generic instructional variables such as questioning strategies, etc. (Slangen et. al., 2011); (3) Knowledge of context and (4) PCK—pedagogical content knowledge. Shulman (1987) defined PCK as the amalgam of all the other domains. The core of PCK is the teacher's understanding of how pupils best learn the concepts of the subject matter. Research shows that teacher training should first focus on the development of teacher's SMK and then develop their PCK (Rohaan, 2012; Shulman, 1987; Slangen et al., 2011). In the current research, the adaptive control system is the concept included in the SMK and PCK that contains structural, procedural and causal knowledge.

The goals of having students develop their conceptual knowledge through experience and active learning, and ensuring they will be able to construct their mental models, require teachers who have the appropriate control mental models as well as the relevant PCK that also includes scaffolding skills.

It is known that K-12 pupils can design, construct and program while working with adaptive control systems. The role of their teachers is to design the learning environment and create the appropriate instruction in ways that enable the pupils to possess the correct control mental models. In order to do this, the teachers themselves must possess the correct mental models of control systems alongside their SMK and PCK.

Based on this literature review, this chapter presents a study that examined the issue of what mental models of control systems student teachers have and then present an instructional unit developed to enhance the construction of correct mental models of self-regulation control processes in technological systems.

9.3 Student Teacher's Mental Models of Control Systems—The Research Description

The following research description contains the research question, method, findings and discussion that will be the basis for the design of the adaptive control instructional unit.

9.3.1 Research Questions

The question in this research is: What are the student teacher's mental models of control processes in control systems?

This question is divided into the following sub-questions:

1. Do student teachers give structural explanations (mentioning operational and control components)?
2. Do they give functional explanations (mentioning processes)?
3. How do they describe the control process?

9.3.2 Method

The research on the student teacher's knowledge of self-regulation control processes prior to learning this subject took place at Beit Berl College, Israel with 37 female undergraduate students attending "Technological Thinking and Robotics", a compulsory course for students in the kindergarten and primary school (K-6) track. These student teachers will teach control systems as part of the mandatory science and technology curriculum for these grades when they graduate.

The course is divided into two main issues, first, technological thinking—design thinking and problem solving, and second, self-regulation control system and robotics. The aims of this course are that student teachers will be able to (a) design—helping student teachers construct their design mental models and (b) analyse, explain, operate and design self-regulation control system based on the mental models of control systems that they have constructed. In the second half of the course, after a short introduction, the student teachers experience the behaviour of the Dash and Dot robot and programme it using the Blockly programming language. This part of the course consists of six 90-min meetings. The course in this format was delivered for six years. As course lecturers, after a few years, we understood that even if the students had fun and enjoyed operating the robots, they were still unable to construct correct mental models of self-regulation control systems.

This dissatisfaction with the student teacher's ability to construct the right mental model after studying this subject for half a semester led us to the conclusion that we need to analyse student teacher's prior mental models and develop an instructional unit that will address their misunderstandings and misconceptions. This research was conducted during the 2018–2019 academic year.

During this course, student teachers need to develop their SMK and PCK of this subject as well as correct perceptions of control systems. In order to develop appropriate instructional unit, a survey of the student teacher's prior mental models was conducted.

In order to answer the research question, participants were asked to describe how an everyday adaptive control system (an automatic door) works, either by sketching or in writing. None of them explained it in writing only, 14% used only sketching without text and 86% used both. In most cases (84%), when the students used both text and sketches, they supported each other. The analysis was based on 37 student teacher's sketches and on 32 texts.

Sketches and texts were analysed according to the following categories and criteria (based on the literature and validated by three judges):

1. First sub-question—Do student teachers give structural explanations? Do they mention components in their explanations? Do they have a full/incomplete/extra structural model? Do they use of OU and CU components (one of the following: for OU undefined/collection/coherent and for CU undefined/collection/control)? (Table 9.1).

2. Second sub-question—Do they give functional explanations? Do they mention aspects of how the system works? Do they describe characteristics of the functional model (one of the following: full/incomplete/extra)? (Table 9.2).
3. Third sub-question—How do they describe the control process? Is their process explanation correct/incorrect, sequential/rule-based? Is their perception of control black box, reactive, switch or control? (Table 9.3).

9.3.3 Findings

The finding will be presented respectively to the sub-research questions: (1) the student teacher's structural explanations; (2) the student teacher's functional explanations; and (3) the student teacher's descriptions of the control process.

(1) Structural descriptions

In testing the students' sketches and textual explanations on how the automatic door works, and analysing them according to the categories and criteria, we understand that:

(a) most mentioned components in their explanations (97% of the sketches and all of them in writing);
(b) most had components missing from their sketches (95%) and in their texts (91%);
(c) only a few mentioned all the components involved in the process (3% by sketch and 6% in writing);
(d) some of the students added extra non-existent components that are not part of this system (14% in sketches and 22% in writing), e.g. signs, bicycles, trees.
(e) In all their sketches student teachers mentioned 71 OU components such as door or sensor, and 68 in their texts, around two components on average for any explanation.
(f) While explaining the CU, only two control components are described in all the sketches and five in all their text explanations (Table 9.1).

The students have a better understanding of the OU components (using on average around two components per sketch and text) than of the CU components, which were mentioned only twice by sketch and five times in text.

It was not possible to analyse the OU and CU configuration from the student teacher's sketches; this was based on their texts only. It was found that most of their OU configurations (that could be analysed as undefined, collection and coherent) were collections (66%) and their CU (that could be analysed as undefined, collection and control) was undefined (84%).

Therefore, in summary, the student teacher's mental models of the structure of the automatic door system contain some components (mostly sensor and door) from the OU, ignoring the CU components. Most student teachers have a partial structure model of the OU configuration and an undefined model of the CU.

Table 9.1 System's structural mental model as presented by sketches and textual explanations

Category	Criteria	Sketch	Text
Structural	Components		
	Mentioned components	97%	100%
	Full Missing Extra	3% 95% 14%	6% 91% 22%
	OU Existing components	71 in all	68 in all
	CU Existing components	2 in all	5 in all
	OU Configuration		
	Undefined Collection Coherent		16% 66% 19%
	CU Configuration Undefined Collection Control		84% 13% 6%

(2) Functional descriptions

The functional mental model answers the question of how the control system works. Sketches did not give the whole picture of the student teacher's functional mental models; however, the text explanations and the integration of both gave a better understanding. The analysis was conducted only on almost half the student teacher's sketches (46%) and on most texts (84%).

Analysis of the functional explanations reveals that many of them present an incomplete model (43% sketches, 75% texts), and only few from both sketches and texts showed full functional models (3%, 9%) (Table 9.2).

The text explanations show how the system works. Two-thirds of the students (66%) explained that the sensor opens the door; others mentioned a sensor operating

Table 9.2 System's functional mental model as presented in sketches and textual explanations (in percentage)

Category	Criteria	Sketch (%)	Text (%)
Functional	Not functional	54	16
	Full Missing Extra	3 43 3	9 75 13
	The process Sensor-doors Sensor-motor Sensor-controller Motor doors Controller-motor Controller-doors Other—doors, info, sensor		66 9 13 6 3 9 19

a motor (9%) and the motor opening the door (6%), a sensor transmitting the information to the controller (13%) and the controller to the motor (3%) or the door (9%). Other student teachers (19%) explained that the door opens by itself.

In summation, the student teacher's functional mental models described both by sketch and by text are mostly incomplete and focus on sensors that operate the doors. The mention of full control processes is very rare.

(3) Control process explanations

The student teacher's explanations of the control process are mostly incorrect, and the process occurring in the CU has incomplete descriptions and missing components (81%). However, half of the existing explanations are "IF...THEN...", for example:

> "**IF** the sensor senses an object or a movement **THEN** the doors open to both sides" (S.A.).

One-third of the students described the process as sequences, for example:

> "a) the sensor identifies movement, b) the information is passed to the motor, c) the motor is activated, d) the doors move in both directions." (G.B).

According to the "sequences of qualitative models of the control system" categories (Mioduser et al., 1996), more than half the student teacher's perceptions are "reactive" (59% in sketches and 66% in writing). Only 27% (sketches) and 16% (texts) have the "black box" model. The "switch" and the "control" models are very rare, as was found in other studies (Landsman, 2019; Mioduser et al., 1996; Rodan, 2016) (see Table 9.3).

The above analyses all draw the same picture of student teacher's mental models of the automatic door as a representation of an adaptive control system. They recognized only few components (doors and sensors) from the OU. Most of them did not mention the CU or its components. More than half the students possess mostly incomplete functional models. Most students had incorrect control mental models, and yet the

Table 9.3 Control system explanations as presented by sketches and textual explanations (in percentage)

Category	Criteria	Sketch	Text
Control models	Correct Incorrect (missing)		13% 81%
	Identifying the process		
	None		16%
	Sequential		34%
	Rules		50%
	Perceptions of control process		
	Black box	27%	16%
	Reactive	59%	66%
	Switch	8%	9%
	Control	3%	9%

models were described by rules and sequences. Students have incomplete structural and functional models, as they lack information about the control process. This also manifests itself as most of them described "reactive" and "black box" models rather than "switch" and "control" models that reflect understanding of control processes. The instructional development is based on those student teacher's mental models that indicate their partial and incorrect SMK and PCK.

In Fig. 9.7 there are four different examples of the student teacher's sketches that explain how the automatic door works. Their analysis, according to the chosen criteria, is below the sketches.

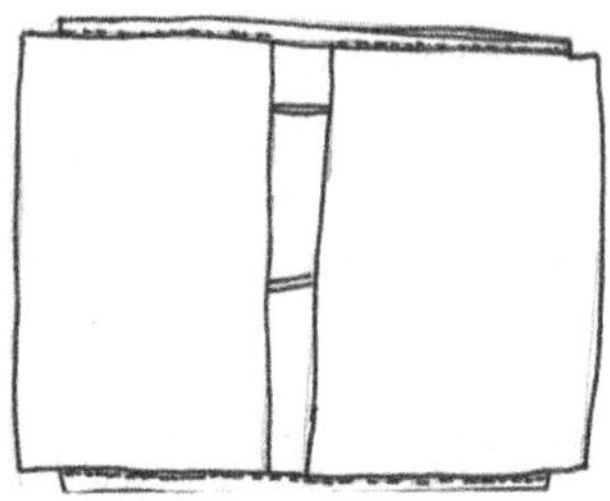 Example no. 1 analysis: Structural: just one component (door), missing, undefined OU, CU. No functional model. Perception of control: 'Black-Box'.	 Example no. 2 analysis: Structural: Two components (door and sensor), missing, OU collection, undefined CU. Functional: using an arrow and person, missing. Perception of control: 'Switch'.
 Example no. 3 analysis: Structural: Two components (door and sensor), missing and extra, OU collection, undefined CU. Functional: using an arrow and person, missing, Perception of control: 'Switch'.	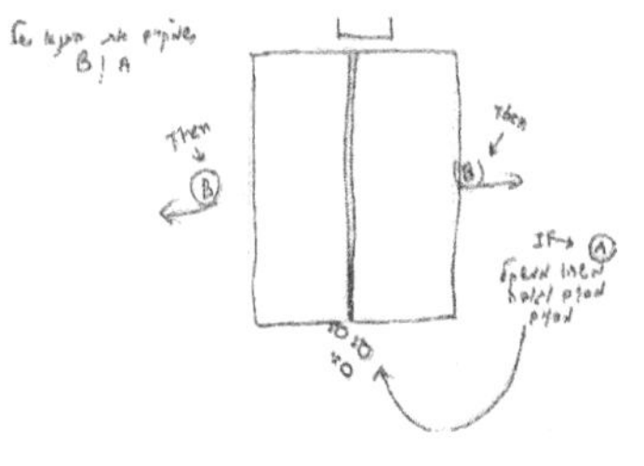 Example no. 4 analysis: Structural: Three components (door, sensor and controller), missing, OU coherent, CU control. Functional: using arrows, full, sensor-> controller-> door. Perception of control: 'Control' and a rule If…Then….

Fig. 9.7 Examples of analysis of student teacher's sketches

9.3.4 Discussion

Most of our everyday life systems and especially control systems are designed as a "black box". People cannot see the system's components, how it works, especially the flow of energy and the flow of information. It is easier to explain those aspects with systems that have visible components (Hallström, 2021).

Although the feedback loop in everyday adaptive control systems is a complex subject, in Israel it is included in the K-6 curriculum. Teachers of this subject must possess accurate mental models of control systems in terms of their structure (their components), behaviour (what is observed) and function (how the system processes the information and produces the desired outputs) (Mioduser et al., 1996. p.387). They need to integrate these models as part of their SMK and PCK in order to scaffold this for their pupils.

Research on learners' technological system concepts and mental models at various ages are uncommon when they are about feedback or adaptive control systems (Hallström, 2021; Hallström & Klasander, 2017; Landsman, 2019; Mioduser et al., 1996; Rodan, 2016). One of the difficulties is that the control processes are invisible and abstract when it comes to transmitting information and decision making with algorithms.

The research described above focused on the question: What are the student teacher's mental models of control processes of everyday control systems? The students' explanations (sketches and text) indicated that for all the three layers (structural, functional and control process), most mental models are partial and include only few OU components and even fewer CU components. These partial models are that "the sensor operates the doors" that indicate that they have "reactive" perspectives (second model in Fig. 9.6) or "when the sensor senses the doors open" that indicate that they have a "black box" perspective (first model in Fig. 9.6).

This research refines the understanding of mental models of control systems by focusing on the student teacher population. The student teachers had only intuitive, partial and inaccurate models that limit their ability to "run" it in their head. Their models are mostly structural, atomistic and partial; they hardly use the abstract aspects of systems, such as flow, boundaries and control, as supported in other research (Hallström, 2021; Hallström & Klasander, 2017; Lind et al., 2019; Mioduser et al., 1996; Norman, 2014; Rodan, 2016). Student teachers rarely expanded their understanding and most of their conceptions remain vague. As we learn from other studies it could be improved with appropriate teaching, experiencing and making the invisible visible (Hallström & Klasander, 2017; Mioduser et al., 1996).

It is crucial that these student teachers have appropriate and accurate mental models of control systems, SMK, CK and PCK that could assist their teaching (Rohaan et al., 2012; Shulman, 1987; Slangen et al., 2011). Studies that used Mioduser et al.'s (1996) models of control systems (Fig. 9.6) indicated that a systematic intervention program that involves building and programming a control system improves learners' mental models. Landsman's (2019) and Rodan's (2016) studies with K-7 pupils showed that the pupils' mental models of control systems improved along the continuum between "switch" and "control". Such scaffolded instruction programs could develop the students' understanding of complex and invisible components and processes, and enable them to construct "runnable" mental models (Hallström & Klasander, 2017; Landsman, 2019; Mioduser, 1996). From their research, Hallström and Klasander (2020) suggested that the instruction could progress from simple to complex systems, from small to large systems and from local to national and global systems. We expect that the instructional unit designed according to the experiences and insights presented here could improve the adult student teacher's prior mental models based only on "reactive" and "black box" perspectives.

9.4 Self-Regulation Control System—Instructional Unit

In order to help student teachers construct their mental models of a self-regulation control system, an instructional unit containing four activities was developed based on the research described above as well as on other studies (Hallström, 2021; Hallström & Klasander, 2017; Koski & De Vries, 2013; Levin & Mioduser, 1998; Mioduser et al., 1996). Each lecturer can choose to use as many activities as they like from the suggested four, considering their students' prior concepts and the time available during the course.

The instructional unit with its various activities implements insights from prior studies and includes a variety of activity types. Learners will interact with the robot, an activity found to enhance generalization, while the lecturer will provide support and scaffolding as needed (Mioduser et al., 1996; Rodan, 2016). The lecturer will choose the technological control systems from simple, small and local, to complex, large and global, respectively (Hallström & Klasander, 2020).

While teaching robotics as part of the "Technological thinking and control" course we used Dash & Dot robots which can be programmed in Blockly. The user can program Dash to move in any direction, change the colours of the lights, play sounds and avoid obstacles. Dash and Dot have some sensors that help them sense their surroundings such as light, sound, angle and ultrasonic. All the sensors, and most of the operators, are hidden inside the robot and are invisible to the users. The student teachers can program the robot to react according to If… Then rules. Dash can dance, move online, play basketball, bowling and more. After six years of teaching with Dash and Dot and observing the student teacher's behaviour, it seems that they enjoyed these lessons very much, but they did not manage to construct proper mental models of self-regulation control systems.

The development of an instructional unit that takes into consideration such research conclusions is needed. The instructional unit presented here is based on making the invisible sub-systems, components and control processes visible. It has four components, from observing the behaviour of the system and analysing it, to watching simulations that show transparent control systems, disassembly and touching the various components and then designing, building and programming the behaviour of a self-regulation LEGO robot; from observing and explaining to manipulating the adaptive control system. We believe that these activities can make the invisible visible. In the four-part unit we emphasize also that teacher support and scaffolding are essential.

The four parts of the instructional unit are:

I. Analysing systems with a variety of self-regulation control units and finding similarities and differences (structural, functional and behaviour/rule use) in the systems.
II. Working with animations and simulations that show transparent control systems, their flow of information and how they really work. Drawing conclusions from these simulations to help mental model construction.
III. Disassembly of everyday life self-regulation control systems such as a pop-up kettle and recognizing the components and their function within the system.
IV. Designing and building adaptive systems (e.g. using LEGO) and programming their behaviour.

Although research has indicated that student teachers who will teach technology and control systems have incomplete mental models and partial SMK and PCK, based on other research (Hallström, 2021; Rodan, 2016), we believe that the systematic instructional unit proposed here could make a real difference.

The instructional unit activities

Activity I. Scaffolding student teachers to analyse various systems with self-regulation control units and find similarities and differences (structural, functional and behaviour/rule use) in the systems. Various systems from the student teachers surroundings, for example traffic lights that give priority when a pedestrian pushes the button, adaptive cruise control in a car, an oven that shuts off when the food is ready, air-conditioning that maintains 25 °C in a room. These systems and others will be examined from the following aspects (a) structural explanations—analysing the system's components—micro understanding; understanding the central components of the control system, their purpose and how they are connected; (b) functional explanations—analysing how the system functions in order to achieve its goals—macro understanding; using the block diagram (Fig. 9.4) to analyse and describe how the system functions; (c) behavioural explanations—what the control system is doing, which includes causal explanations by rules.

In this activity, student teachers will:

(a) choose an adaptive control system from a list.
(b) search and study how it works
(c) find out and describe the structure of the system
(d) find out and describe the functions of the system—could be done (recommended) with a block diagram.
(e) explain the behaviour of the system
(f) explain the rules of the system's If... Then...decision making
(g) compare the similarities and the differences between control systems in groups

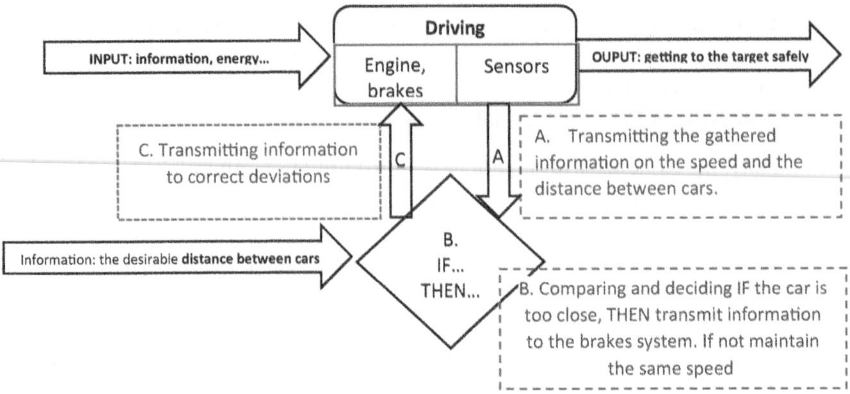

Fig. 9.8 A block diagram of adaptive cruise control

The adaptive cruise control, for example, can adjust speed in order to maintain a proper and legal distance between vehicles in the same lane. The OU components are sensors, the engine and the brake system. The control system components are the controller. The functional explanation: the sensors that collect information about the surrounding cars, pedestrians or other objects in front of the car send signals to the controller which, according to its algorithm, decides the speed of the car and accordingly sends the signal/information to the car engine or the braking system. It can be described by a block diagram (Fig. 9.8).

The control rule will be: IF the car is getting closer to another car or other obstacle (getting closer is according to the comparison between the information about the desirable distance and the information received from the sensors), THEN transmit information to the engine to stop working and to the brake system to operate. IF it is the right distance, THEN the information transmitted to the engine is to maintain the same speed.

Each student teacher will analyse a self-regulation system as described above. In teams, they will compare the systems as presented in Table 9.4. From the table they will find out that the similarities would be the flow of information, the use of sensors, operators and rules for decision making. The differences could be the specific component—sensors and operators, and different rules. This comparison could assist generalization and reflection, both of which enhance the construction of proper mental models of adaptive control systems.

Table 9.4 Comparison between systems

Systems	Structural	Functional	Behavioural
System 1			
System 2			

The lecturer will be there to scaffold the learning process in a way that enables the student teachers to construct their mental models of adaptive, self-regulation control systems.

Activity II. Using animations or simulations of transparent control systems that show the components (structure), their functioning and behaviour and how each of them really works. Observing these transparent systems makes it possible to identify the components and the interrelations between them (structural and micro aspects), to find out how the system functions and elicit the control rules. Using a block diagram is one way of analysing the system, but the student teacher could be asked to explain it also by sketching or in writing. This learning process could help learners understand better and even construct a generalization of a control system as a step on the way to constructing their mental model.

Each of the student teachers will analyse a control system and, in groups, they will discover the similarities and differences between them (as described in activity I) in order to create a generalization of adaptive control systems. This generalization could help mental model construction.

Activity III. Disassembly of everyday life self-regulation control systems such as a pop-up kettle, a flush toilet tank, etc. and recognize the components, their function and the adaptive control system rules. Sometimes it is not easy to understand how the system works, or what the function of each component is, so we recommend integrating this activity with one of the previous ones (activities I or II). In this activity, the student teacher will describe the system's structure, function, behaviour and the control rule by sketching, a written description or in a block diagram (or both).

Activity IV. Designing and making adaptive systems (e.g. using LEGO) and programming their behaviour. The Dash & Dot robots are like a black box to the users. They know that they can use light, sound and motors operators; they also know about the sound, light and other sensors and they learn how to program it. For example IF the sound is XX, THEN go left at a speed of XX, etc.

In order to look inside the black box and feel and touch the components, we suggest using the LEGO SPIKE robot (the LEGO MINDSTORM is too complicated for this specific population and for the limited course time).

The student teachers will be asked to design and build the robot, its structure, behaviour and function, from SPIKE building blocks. They will use sensors and operators like lights and motors and program its behaviour. By doing that, student teachers can feel and understand the relationship between the components and construct their mental model step-by-step in layers (Norman, 1983). While programming their robot, another layer of this mental model is constructed. After designing, building and programming, the student teachers will reflect on this process by describing their system in writing and/or sketching and/or a block diagram.

Lecturers can use all or some of these activities to scaffold and enhance student teacher's construction of their adaptive control systems mental models according to their prior concepts, SMK and PCK, and the time constraints of the course.

Before and after experimenting with these activities, student teachers could be asked to explain how the automatic door or the adaptive traffic light works in order to find out if their mental models have changed following experiencing the learning activities and if they now possess the right concepts and appropriate runnable mental models that could help them understand everyday life self-regulation control systems.

9.5 Conclusions

We are surrounded by smart, adaptive, self-regulated control systems. As human beings we need to be technologically literate and understand how they work in order to use, change, troubleshoot and even design new ones. Most of these control systems are black boxes because their processes are invisible and abstract.

These current study findings about student teacher's mental models of adaptive control systems reinforce previous studies on this subject for various age groups. They all indicated that students of various ages, student teachers and even active teachers have partial mental models of control from the three aspects: structural, functional and behavioural. We used these findings as a reference point to develop an instructional unit containing four activities emphasizing the following aspects:

- Finding and understanding similarities and differences
- Authenticity—choosing well-known self-regulation systems from the student teacher's surroundings.
- Using block diagram analysis
- Making the invisible visible and comprehensible
- The control system as flow of information
- Using (interact, design, make and compute) an adaptive system such as a robot.
- Analysis, synthesis, creation and evaluation.
- Interacting with simple to complex, from small to large and from local to global systems.
- Scaffolding and support
- Generalization and abstraction

Asking the student teachers to explain how an adaptive control system works before and after experiencing the four activities could indicate the effectiveness of the instructional unit. Student teachers trying out a smart control system will be able to describe their mental models of smart systems as "switch" or even "control" and less as "reactive" and "black box" (Fig. 9.6). Further research is needed to examine the effectiveness of the unit.

References

Barak, M. (2018). Teaching electronics: From building circuits to systems thinking and programming. *Handbook of technology education*, 337–360. Springer. https://doi.org/10.1007/987-3-319-44687-5

Dagan, O. (2021). Student teacher's mental models of everyday control systems—The sensors that operate the door. *Techne Series—Research in Sloyd Education and Craft Science A, 28*(2), 62–71. Retrieved from https://journals.oslomet.no/index.php/techneA/article/view/4380

DeKleer, J., & Brown, J. S. (1981). Mental models of physical mechanisms and their acquisition. In: J. R. Anderson (Ed.), *Cognitive skills and their acquisition*. Lawrence Erlbaum.

Grossman, P. L. (1990). *The making of a teacher: Teacher knowledge and teacher education*. Teachers College Press.

Hallström, J. (2021). Evaluating an intervention to improve secondary pre-service teachers' conceptions of feedback in technological systems. *Techne Series—Research in Sloyd Education and Craft Science A, 28*(2), 204–212. Retrieved from https://journals.oslomet.no/index.php/techneA/article/view/4288

Hallström, J., & Klasander, C. (2020). Making the invisible visible: Pedagogies related to teaching and learning about technological systems. In P.J. Williams & D. Barlex (Eds.), *Pedagogy for technology education in secondary schools* (pp. 65–82). Springer. https://doi.org/10.1007/987-3-030-41548-8

Hallström, J., & Klasander, C. (2017). Visible parts, invisible whole: Swedish technology student teachers' conceptions about technological systems. *International Journal of Technology and Design Education, 27*(3), 387–405. https://doi.org/10.1007/s10798-016-9356-1

Hegarty, M. (1991). Knowledge and processes in mechanical problem solving. In R. J. Sternberg & P. A. Frensch (Eds.), *Complex problem solving* (pp. 253–285). Lawrence Erlbaum.

Jones, A. (2003). The development of a national curriculum in technology for New Zealand. *International Journal of Technology and Design Education, 13*, 83–99. https://doi.org/10.1023/A:1022355410425

Kieras, D. E. (1988). What mental model should be taught: Choosing instructional content for complex engineered systems. In J. Psotka, L. D. Massey, & S. A. Mutter (Eds.), *Intelligent tutoring systems—Lessons learned* (pp. 85–111). Lawrence Erlbaum.

Kieras, D. E., & Bovair, S. (1984). The role of a mental model in learning to operate a device. *Cognitive Science, 8*, 255–273.

Kipperman, D., Dagan, O., & Shechter, G. (2000). *Technological systems* (textbook).ORT Israel. Moshinski R&D Center (pp.175–204) in Hebrew.

Kirkpatrick, D. (1996). Great ideas revisited. *Training and Development, 50*(1), 54–59.

Koski, M. I., & de Vries, M. (2013). An exploratory study on how primary pupils approach systems. *International Journal of Technology and Design Education, 23*(4), 8. https://doi.org/10.1007/s10798-013-9234-z

Landsman, L. (2019). *Preschool children perceptions of control systems work following daily usage and following working experience with control systems*. MA thesis. Supervised by D. Mioduser, In Hebrew.

Levin, I., & Mioduser, D. (1996). A multiple-constructs framework for teaching control concepts. *IEEE Transactions on Education, 39*(4), 488–496. https://doi.org/10.1109/13.544802

Lind, J., Pelger, S., & Jakobsson, A. (2019). Students' ideas about technological systems interacting with human needs. *International Journal of Technology and Design Education, 29*(2), 263–282. https://doi.org/10.1007/s10798-018-9449-0

Mioduser, D. (1998). Framework for the study of cognitive and curricular issues of technological problem solving. *International Journal of Technology and Design Education, 8*(2), 167–184. https://doi.org/10.1023/A:1008824125352

Mioduser, D., Venezky, R. L., & Gong, B. (1996). Students' perceptions and designs of simple control systems. *Computers in Human Behavior, 12*(3), 363–388. https://doi.org/10.1016/0747-5632(96)00014-3

Mioduser, D., & Dagan, O. (2007). The effect of alternative approaches to design instruction (structural or functional) on students' mental models of technological design processes. *International Journal of Technology and Design Education, 17*(2), 135–148. https://doi.org/10.1007/s10798-006-0004-z

Norman, D. A. (1983). Some observations on mental models. In D. Genter & A. L. Stevens (Eds.), *Mental models* (pp. 7–14). Lawrence Erlbaum.

Norman, D. A. (2014). Some observations on mental models. *Mental models* (pp. 15–22). Psychology Press.

Rodan, I. (2016). *Development of mental models of control systems following the experience in programming robots among 7-year-old children.* MA thesis. Supervised by D. Mioduser, In Hebrew.

Rohaan, E. J., Taconis, R., & Jochems, W. M. (2012). Analysing teacher knowledge for technology education in primary schools. *International Journal of Technology and Design Education, 22*(3), 271–280. https://doi.org/10.1007/s10798-010-9147-z

Rzevski, G. (1995, August). Intelligent systems: Issues and trends. In *International Conference on Intelligent Manufacturing* (Vol. 2620, pp. 14–23). International Society for Optics and Photonics. https://doi.org/10.1117/12.217479

Schooner, P., Klasander, C., & Hallström, J. (2015). Swedish teachers' views of assessing technological systems in compulsory school. In M. Chatoney. (Ed.), *PATT 29. Approaches in Design and Technology Education,* Presses Universitaires de Provence.

Shulman, L. (1987). Knowledge and teaching: Foundations of the new reform. *Harvard Educational Review, 57*(1), 1–23. https://doi.org/10.17763/haer.57.1.j463w79r56455411

Slangen, L., van Keulen, H., & Gravemeijer, K. (2011). Preparing teachers to teach robotics in primary schools. In *Professional development for primary teachers in science and technology.* Brill. Retrieved May 23, 2021, from https://brill.com/view/book/edcoll/9789460917134/BP000015.xml

Svensson, M., & Klasander, C. (2012). Teacher's professional growth in planning and teaching technological systems. In *Technology Education Research Conference, Surfers Paradise, Australia.*

White, B. G., & Frederiksen, J. R. (1986). Progression of quantitative models as a foundation for intelligent learning environments, Technical Report # 6277, Bolt, Beranek, & Newman. Retrieved May 23, 2021, from https://files.eric.ed.gov/fulltext/ED275300.pdf

Osnat Dagan is PhD in technology education from School of Education at Tel-Aviv University, Israel. Her PhD was focused on problem solving in design and technology. Osnat currently works as a lecturer and as the head of ICT academic unit at Beit Berl College. Her academic interests are STEM; design and technology; the use of innovation pedagogy with ICT technology such as virtual worlds and chatbots; develop thinking skills and problem solving skills of learners using constructionist methods and teacher professional development. Her two latest publications are: Dagan, O., (2020). STEM education in the secondary school in Israel. In D. Barlex & F. Banks (Eds.), Teaching STEM in the secondary school. Routledge. UK; Dagan, O. (2021). Student Teacher's Mental Models of Everyday Control Systems—The Sensors that Operate the Door. Techne Series—Research in Sloyd Education and Craft Science A, 28(2), 62–71. https://journals.oslomet.no/index.php/techneA/article/view/4380

Chapter 10
Learning and Teaching with a Systemic Approach: Using Functional Analysis in Technology Education for All

Marjolaine Chatoney and Fabrice Gunther

Abstract The purpose of this chapter is to show that learning technical systems, with tools designed by engineers, allow students and teachers to better understand the multi-technological and environmental complexity affecting the technological choices of the object or system concerned. A systemic approach makes it possible to consider the complexity of systems. To do that, engineers use tools that consist of studying systems function by function. This method of reasoning is called a functional approach; the tool associated is called functional analysis. Both work together and favour a multidisciplinary approach that integrates and connects different subjects, which are interwoven in the technical system. They decompartmentalize functions. Each little part of the technical system makes sense without forgetting the system as a whole. The functional approach favoured is using these tools and provides a systemic and complete analysis of a system with a holistic approach. It allows users to perceive all the complexity of the interactions that there are between people, systems and the environment of use of these systems, that is why it is a fundamental part of the technology curriculum. From the point of view of teaching and learning in technology education, we propose a systemic approach coupled with the tool of functional analysis. Both work as a frame for teacher, oriented towards objectives that make sense for the acquisition of knowledge for pupils. The students can connect the prescribed activity with the aim they have to reach and are thus more efficient. The efficiency and the need to give meaning are linked, if the activity makes sense, the learners are much more efficient. In this chapter, we will present two results of research. The first one informs and characterizes the uses of functional analysis in practices reported by technology teachers in middle school. It also informs what students learn and can do with their learning when they are taught techniques from functional analysis. The second research results measure the effect of transferred technical systems learning in physical science problem solving in high school. It shows that students who have learned the systemic approach coupled with the tool of functional analysis develop a systemic reasoning benefit, which is useful to understand other technological systems.

M. Chatoney · F. Gunther (✉)
Aix Marseille University, 58 Boulevard Charles Livon, Marseille 13007, France
e-mail: fabrice.gunther@laposte.net

J. Hallström and P. J. Williams (eds.), *Teaching and Learning about Technological Systems*, Contemporary Issues in Technology Education,
https://doi.org/10.1007/978-981-16-7719-9_10

10.1 Introduction

The issue of teaching–learning in technology education is directly related to the question of the reference of knowledge to be taught. This reference to knowledge varies from country to country. In France, the knowledge to be taught refers to knowledge about entrepreneurial practices of object production and technical systems.

As the curriculum indicates (BOEN, 2001), a systems approach, which forms the basis of teaching, is confirmed as an important part of the curriculum of technology (here, technology is a subject in general education, as TE). Several components of the systemic approach, functional, structural and behavioural, allow one to determine how a system should be perceived and used. Among other characteristics, this teaching structure is presumed to be consistent with current engineering practices used in industry. Therefore, we can expect teachers to use the systemic approach to build both knowledge about technical systems, and at the same time, methods of analysis of companies transposed to the classroom.

The purpose of this chapter is to show that learning technical systems, with tools designed by engineers, allow students and teachers to better understand the multi-technological and environmental complexity affecting the technological choices of the object or system concerned. This chapter reports on a first research study which informs and characterizes the uses of functional analysis in practices reported by technology teachers in middle school. In a second study, it shows the effect of transferred technical systems learning in physical science problem solving in high school.

Analysis of technical objects and systems plays a central role in the processes of teaching/learning of technology. These analyses appear at all levels of education, from kindergarten to high school. To define how to analyse a system, we refer to what we understand from the work of Le Moigne (1999) who contrasts two types of approaches: analytical and systemic. He assumes that while the most common approaches are analytical, they do not reflect the complex situation of a system. The systemic approach considers the function and its actions, which are not done by the analytical form that describes a closed system and reasons on a single criterion. Le Moigne also criticizes the education system for choosing the path of simplicity by teaching only analytical models, more easily accessible than systemic models. Also, for Le Moigne, decision-making processes do not appear in analytical models, whereas systemic modelling incorporates this decision-making factor and may result from the modelling of cognitive problem-solving processes, which we will discuss later in the chapter.

10.2 System, Systemic Approach, Systemic-Functional Analysis

10.2.1 Origin, Link to Engineering and Design Sciences

Systemic approach (SA) appeared in the field of biology in 1920 in EU, the war caused its transfer to the United States in 1940. It was originally linked to the theories of the scientific organization of labour promoted by Le Chatelier, Taylor and Ford, whose objective was to move from a method of artisanal production to industrial production (Aïm, 2013). It therefore comes from the world of industrial production. It was popularized in France, by de Rosnay (1977) in biology, Morin (1990) in the social field, by Le Moigne (1999) and by Simon (1996) in the field of artificial intelligence and Inerheld-Cellerier (1992) in cognitive sciences. SA, which is defined as a dynamic and complex set of interactions, is a structured functional unit of interdependent elements, the whole of which is greater than the sum of its parts (Von Bertalanffy, 1968). It is, for these reasons, more adapted to the needs of today's (systemic) world than Cartesian (linear) logic. It characterizes systems by specifying their limitations, internal and external relationships, structure and new laws; it makes the company's socio-organizational functionality manifest and makes it more fluid (Scaravetti, 2004).

Industrial technology classifies under the term "Systemic and Functional Analysis" (Howard, 2007) different methods that allow understanding of the systemic dimension of a technical system (Zehtaban and Roller, 2012) from the point of view of its functionality. Each of these methods is associated with a tool. This is the case with applied methods to enterprise called structured analysis and technical design (SADT), functional analysis system technique (FAST); Application aux Techniques d'Entreprise (APTE®), CDCF (an acronym for functional charges), etc.

10.2.2 Methods and Tools to Understand the Complexity of Systems

"Systemic and Functional Analysis" constitutes a set of techniques used in the creation or improvement phase of a product to identify real needs, quantify them, define real problems and grasp what is important to obtain. It is an approach using different tools to characterize a product in the form of graphs or diagrams. It is useful in the design or improvement phases of the project. Functional analysis is part of the technical language. It is a semiotic register that complements other semiotic registers: natural language, graphic representation, writing, drawing, etc. We call these techniques "functional analysis".

SADT and FAST methods are associated with graphic tools (diagrams, graphs) and textual contents (specifications). All these tools are symbolic representations of the system that allows the system to be communicated and understood using a common language (Scaravetti, ibid.). In other words, functional analysis presents itself as a set of methods and tools most often used in the design or improvement phase of industrial products. These methods and tools are useful in helping to identify, quantify and solve problems related to needs. Moreover, this methodology, as Vernat (2004) points out, serves both in the formalization phase of constraints and in the writing of parsimonious and accurate models. From this perspective, the functional analysis that combines function and structural elements (Scaravetti, ibid.) is a tool to understand technological systems.

10.2.3 Why Systemic Approaches in Education?

The need is to consider technical artefacts as systems rather than simple objects that can be taken apart and which are without any clear boundaries (Deforge, 1993; Gilles, 1978). The systemic approach also led to a new way of designing technological education (Barak, 1990, 2007; Brown et al., 1989; de Vries, 2005; Dorst, 2006; Dubois & Gartiser, 2005).

The systemic approach is also a significant change that most likely has a positive effect on students' ability to deal with problems in a more comprehensive and structured manner. It is introducing another way of reasoning in addition to the analytical approaches taught for centuries in science (Andréucci et al., 2010).

When teachers or students use these tools, teaching is referenced, and learning makes sense to students (Ginestié, 2000). Empirical studies conducted in primary school (Chatoney, 2003, 2005, 2013) indicate that the functional analysis is a structuring instrument that makes technology education sustainable in primary school. In other words, the functional approach allows, among other things, to include technological education in the context of human activities of the production of technical objects and no longer as simply the application of science.

However, the implementation of functional analysis from the industrial world reveals several problems: a problem of transposition of industrial tools in the field of teaching and learning, in particular on the knowledge involved and references (Graube et al., 2003), and a problem of transformation into instruments for teaching.

10.3 Systemic and Functional Analysis in French Technology Education

10.3.1 Functional Analysis Dedicated to the Teaching of Technology Education

Technology education (TE) is a general education subject in France. This subject is compulsory for all students from 3 to 15 years old. After that, technology education becomes optional until the "bachelor" examination (17–18 years old). From 15 to 18 years old, technology education is taught in the course "Introduction to Engineering Sciences" (ISI) an option for 15–16 year olds, and in the "Engineering Sciences" (SI) option for 16–18 year olds. Not all students choose technology, they can choose other options such as: art, foreign and classic languages, economic and social sciences, management information and communication, physics–chemistry laboratory, physical education and sports activities. The choice of these options depends largely on the type of *baccalaureate* aimed for at the end of year 13. In the history of the teaching of technology, the objectives have not always appeared to be essential and have not been perceived as very stable.

A study of official curricula and textbooks indicates how functional analysis can be studied. Educational choices and the epistemological environment are indicators of how knowledge is transmitted (Johsua & Dupin, 1993). In the case of functional analysis, transposition can be done with at least two different approaches. The first is an indication of the main concepts such as interrelationships in a system, the relationship of man to the system. Chatoney (2003) explains that that comes from conceptualizing the attributes of generic acceptance and selecting information. This approach is intended for teachers. Another more pragmatic approach considers teaching tools from the description of an object, it is a utilitarian model that is proposed for the learner. It is a question of providing a representation that can assist in conceptualization (Martinand, 2003) and that is inherent in practice (Lebahar, 2006). In middle school, technology programs also include an economic aspect and always have knowledge to acquire that relates to the cost of a technical object or materials used. This input by value is associated with functional analysis and analysis of value (Jouineau, 1982). It is through proper functional analysis that the costs of designing and manufacturing a product can be reduced. Here, we find some of the systemic aspects presented previously in the sense of an holistic approach.

10.3.2 A Tool for the Teacher in Relation to the Curriculum

What does it mean that it is a tool (consistent with the heading) for teaching? Using a systemic approach to teaching can improve the efficiency of the learning and teaching process. Systemic analysis is an instrument in the sense of Rabardel (1995). The tool allows teachers to explain a system to students and so becomes an instrument. It allows them to deepen learning by focusing on one or more functions without losing the purpose of the system studied, in other words, without having to study all the functions (Chatoney, 2003).

The functional analysis is a useful tool for perceiving complexity. But it also has other functions that go far beyond the technical aspects; it is also an instrument of teaching. In this case, for us, it is an instrument that organizes the teacher's work, it refers to knowledge and, in fact, allows the teachings to be organized. This approach is updated by studying how teachers understand and use the tools of functional analysis. It is a question of seeing their relationship with this method, and how they use these tools in their lessons or school sequences.

In technological education, how important and valid is the functional analysis? The evidence provided allows us to say that it is an industry-based method that lays a theoretical foundation. It is universal, validated and widely applied. It has different modes of communication; diagrams, writings… which are symbolic languages and can benefit students. Following the field observation, we realized that there is a great heterogeneity in teacher's practices. Some teachers use these approaches as a tool of thinking and reasoning about systems, others do not use it. The curriculum indicates that this tool must be taught. In general TE, the French approach is to incorporate systemic reasoning in order to:

(1) Ensure that students, from an early age, experience this reasoning through the teacher when they are in a situation of studying technical systems.
(2) Ensure that students use the tool to reason and understand the complexity of the technical systems that they have to study.

From the perspective of further studies in high school, this "elementary brick" is akin to a "common foundation" of technological education necessary to analyse technological objects and systems.

10.3.3 What Kind of Learning is Effective? (Making Sense for Students)

Functional analysis as an instrument is not only effective for teaching but also for learning. In France, the middle school is where the beginnings of functional analysis should be taught. It is integrated into official curricula and leads students to an understanding of the complexity of technical systems. The effectiveness of the teaching of functional analysis is addressed through the study of systems. The final perspective is to show the place of learning a specific language in the construction of technical thought. The relationship with the technical object is discussed in terms of its cost, marketing or value to the user or buyer. These functions are integrated into the understanding of how the object works, its design, its manufacture and the materials used. It is this combination of purely technical and social, economic or artistic aspects that will bring the student to a certain level of thinking. We will look at the student's representations and how they appropriate the tool for themselves. Its effectiveness will be studied by a comparison of results.

The middle school in France, corresponds to an age group of 11 to 15 years old where students are at a stage of development sufficient to construct these concepts. They are gradually evolving towards adult status. At this age, students have sufficient knowledge to address skills that correspond to this status. They are still receptive to concepts that have not been addressed and that are likely to give them a certain view of the world. For these reasons, we chose to conduct a study at this level of education. Simondon (2012, 309–311) stresses that what comes from a technical field must remain technology-based and notes that technical patterns and processes derived from reflexive thinking must be applied to technical reality. It is not clear that technical reality can be known through concepts. He points out that to acquire technical knowledge, it is necessary for the human being to be put in a situation where it is necessary to solve a real problem that makes sense. We will keep in mind that we need to stay in the field of technology and as close as possible to real situations. We choose to study the teaching process "in situ" of the classrooms—learning with all the problems that this poses methodologically: agreement with teachers, their directions and disruption of the annual schedule. We thus observe reality and not a model of reality. The reality of the classroom is complex, but our research focuses on this reality with an authentic teacher and students. In relation to functional analysis, we are entitled to ask ourselves how this industrial instrument can contribute to the acquisition of knowledge that allows us to analyse and understand technical systems. To answer this question, we assume that functional analysis is not only an instrument used by technology teachers, but also that it leads students to an understanding of the complexity of technical systems.

10.4 Example 1: Studying the Effectiveness of Functional Analysis Tools in Teaching and Learning Technical Systems in Middle School

Our theoretical framework has helped to explain the complexity in which teaching–learning processes are put in place. To try to analyse this complexity, our methodology relies on the elements presented, instrumental genesis, cognitive processes, in the actors of teaching and learning, with both teachers and students. The tools derived from functional analysis, objects of teaching, are used by the teachers like any other material in the classroom and are perceived as an artefact. Based on research performed by Gunther (2016), this example presents the results and methodology used to learn more about teachers' use of functional analysis and what students learn and can do when they are taught techniques from functional analysis.

10.4.1 Method

Teaching:

In technological education, a commonly used representational technique is based on functional analysis. Depending on their skills, teachers rely on this technique and may or may not use functional analysis. To find out how teachers use functional analysis and what their classroom practices are, we chose to perform a survey. The survey consists of an online questionnaire submitted to about 1200 teachers with a response rate of roughly 11% to determine the reported practices. The goal is to find out if teachers are using functional analysis and if so, in what ways. To do this, we asked different questions that allow us to identify at the same time: their knowledge of the tools of functional analysis, the state of their practices about these tools, certain personal characteristics of the teacher: age, training and seniority in the profession. The aim of the interviews is to know the teacher's reactions to documents concerning the study of systems and to analyse the responses through speeches.

Learning:

To deal with how students can take ownership of the functional analysis system technique (FAST) tool of functional analysis and approach its effectiveness, we carried out an experiment with students in teaching situations during a technology course at a middle school. The experimentation setup allows quantitative and qualitative analysis of the learning activity. It is designed to assess student skills and to see if there is a trigger that allows students to reinvest those skills in a particular situation.

Quantitative analysis is based on exercises with written productions by students. We note the number and pertinence of the responses given by the students (sort of the technical manipulations, description of the system).

Qualitative analysis consists of an operative phase with a manipulation of the 3D printer and a fabrication carried out by pairs of students. We note the technical capacity of the students to produce a form (number of technical operations, results).

10.4.2 Results

The aim of the study is to know and characterize the uses of functional analysis in practices reported by technology teachers in French middle schools. Based on a questionnaire, we analyse teachers' knowledge and practices on certain tools of functional analysis. We show that teachers, even though they use it very widely in middle schools, do not always say they are in control of it perfectly. By categorizing teachers and their responses, we try to find clues and highlight the main difficulties that prevent functional analysis from being a generic method of technological education for all grade levels.

- Quantitative treatment of teaching

Table 10.1 Number of tools known by teachers

Number of known tools out of 4	0	1	2	3	4 and more
% of teachers (%)	14	11	12	17	46

Table 10.2 Teacher's use of tools

Type of tools	Needs chart	Inter-actors graph	Functional specifications	FAST
% of teachers who uses this tool (%)	56	59	77	57

At first, we looked at how teachers use functional analysis. Then, second time, we sought to know the actual practices of teachers when using the tools of functional analysis. The following table shows the number of tools known to teachers among the four most common in functional analysis (needs diagram, inter-actor graph, functional specifications, FAST). Each one of them appears at different phases of the definition of a technical object: to define the needs. to analyse the interactions with the external environment, to specify contractually the constraints and to adapt solutions at the different functions (Table 10.1).

This table indicates that only 46% of the teachers know the tools that allow an analysis of a system. A total of 17% know at least three tools. The other is adding those who do not know enough tools to study systems (14 + 11 + 12%) differently or do not study them and get far away from the prescription.

- The type of tools used by teachers:

The following table represents the percentage of teachers who use different tools (Table 10.2).

The tools, most to less used, are in order the functional specifications (77%), the inter-actor graph (59%), the FAST diagram and then the needs diagram (57 and 56% of users).

- Type of tools used by teachers according to their teaching practices:

The following table represents the distribution of the number of teachers according to the tools of the functional analysis that they use following four pedagogical modalities that are frequently used in technology education. It highlights the link between the pedagogical modality adopted and the type of tool used.

- Pedagogical modality 1: study of existing systems (systems are presented to the student).
- Pedagogical modality 2: system design (systems do not yet exist, they are designed by students).
- Pedagogical modality 3: modality 1 and modality 2.
- Pedagogical modality 4: Teachers do not use functional analysis (Table 10.3).

Table 10.3 Allocation of tools by teaching methods

	% of teachers (%)	Needs chart (%)	Inter-actors graph (%)	Functional specification (%)	FAST (%)
Modality 1	32	34	34	34	34
Modality 2	35	34	33	33	32
Modality 3	25	32	33	33	34
Modality 4	8				

This table shows that 32% of teachers use functional analysis during the study of existing systems. Of the 32% who use modality 1, four tools are used as well as each other (34% for each tool). A total of 35% of teachers use functional analysis in system design. Of the 35% of teachers who use pedagogical modality 2, four tools are used in almost the same proportions, with a slight increase for the needs chart and a small decrease for the FAST diagram. A total of 25% of teachers use pedagogical modality 3, e.g., 25% of teachers use functional analysis both to study presented systems or in system design. Again the four tools are almost as used, but this time with a slight decrease for the needs chart and a slight increase for the FAST. 8% of teachers do not use functional analysis. Even if the tool is known, its use in the classroom is not related to a particular modality. Teachers use functional analysis in all cases. The distribution of tools by study modality is almost identical, one third in design, one third in study and one third for both. The tools of functional analysis are useful for teachers to design and study technical systems. APTE and FAST tools designed for design are also used to decrypt existing systems. For example, the needs chart from APTE focuses the purpose of a technical object, and the FAST diagram is useful to perceive in one table the different functions with an adapted technical solution for each of them.

- Purpose for using the tool:

The table below presents the distribution of teachers by educational objective: understanding functional analysis, understanding the system and understanding both functional analysis and the system (Table 10.4).

Half of the teachers (52%) aim to develop students understanding of both the functional analysis tool and the system studied. A total of 26% of teachers seek to convey the techniques of functional analysis and do not attribute importance to the system that is the learning medium. A total of 22% aimed at understanding the technical system and not the logic of the tool that allows these systems to be

Table 10.4 Distribution of pedagogical concerns

Objective	Understanding functional analysis	Understanding the system	Understand functional analysis and the system
% of teacher (%)	26	22	52

understood. From the epistemological posture of teachers, this reflects their desire to focus knowledge on both the tools and understanding of the system.

- Quantitative results on the order of operations in using a 3D printer by students

The interest in this exercise is to be able to analyse whether students have a comprehensive and functional view of the system. The task of putting in order the operations that lead to the implementation of the 3D printer system requires understanding of all the components and structuring a logic of interaction between the different elements. A sequential temporal logic must be considered, and it complements functional analysis and differs from a material approach. For this exercise, it is recalled, and our sample is divided into two groups:

- initiated to functional analysis noted AF (85 students);
- non-initiated noted NAF (70 students).

What we seek to define and understand lies in the comparison between the two groups rather than in the intrinsic responses given by the students.

However, the students had to build a ball with the 3D printer which was intentionally uncalibrated. They must put in the right order the eight different operations needed to recalibrate and run the printer otherwise the production would fail (Fig. 10.1).

In the NAF group, 5 out of 70 students (7%) gave a correct answer. In the AF group: 10 out of 85 students (11%) gave a correct answer. This little increase shows a contribution of the functional analysis. It must be considered that a large number of students have made mistakes. For this, we will analyse in particular the results of these students.

If the student put an operation in the wrong place compared to the correct order, it is counted as one error. A maximum of eight errors can be made. We then counted the number of errors made by each student and calculated the error rate for each group (number of errors divided by the number of students in the group). The results are in the following table (Table 10.5).

This error rate is much same for students initiated into functional analysis and uninitiated, uninitiated students make an average of 2.87 errors and those who are

Fig. 10.1 Results of productions

Table 10.5 Exercise 1—Average number of errors

	Uninitiated NAF group	Initiated AF group
Error rate/student	2.87	2.90

Table 10.6 Error rate based on general level

Level / Group	A (Students with good scholar results)	B (Students with middle achieving scholar results)	C (Students with low scholar results)
NAF	2.47	2.85	3.26
AF	2.82	2.90	3.06

initiated 2.9 errors out of a maximum possible total of 8. There are few differences between the two groups when considering our total sample of 155 students (less than 0.4%). If we detail these results by categorizing in the way that for the entry test (categories A, B and C based on the general averages grades of students over the year), we get the following table (Table 10.6):

It is found that for the group who were uninitiated to functional analysis, the gap between students who have a good level and those who have a level below the average is almost 10% over the eight errors possible, and the rates vary from 2.47 to 3.26. For students who are initiated into the functional analysis, this gap is only about 3%, rates range from 2.82 to 3.06. It appears that the introduction to functional analysis has led to a homogenization of the levels, if the students in difficulty are slightly more efficient, the students who have more school facilities would then be less efficient. This last finding may sound strange. But if we look in detail at the results of these students, we see that a few of them (3 out of 28) did not finish this exercise, the non-answers being counted as errors. This was not the case for NAF students, and in this group, all students who did not complete the exercise, i.e., 5, belong to category C. Those in the AF group who did not complete the exercise on time are as follows: as noted above three in Category A, two in Category B and three in Category C. It may therefore be thought that the need to analyse the system in its all complexity may have taken longer, including for some less troubled students. Even for them, this is not a usual and scholar way of thinking, and they are not used to this and need time to adapt themselves. This may explain the small change in the error rate between the two groups.

- Qualitative results on learning:

One can, through linguistic considerations, distinguish between notional and conceptual levels. Our intention is to understand how systemic concepts can emerge, and this through what linguistic status. If students have appointed entities using functions instead of technical elements, then a transfer of the concept can be considered. If of course no entity is listed, it is impossible to argue that this transfer is made. So, we studied a total of six student pairs. Three of them had a specific lesson with some examples about how to analyse a system with the FAST technique as an introduction to the functional analysis, the other three pairs did not follow this initiation. For each modality, teaching with functional analysis and teaching without functional analysis, we have trained, in view of the results of quantitative experimentation, a pair of students with "weak" scholar results, a pair of students with "average" results and a pair with "good" results. Only the weakest students who did not follow the teaching of functional analysis failed in the manufacture of a ball with the printer. The first finding is that manufacturing is quite feasible. This time, it is noted that the students of the AF pair behave, at all school levels, better than the NAF pairs. Only the pair of the best NAF group students made no mistake. This means that the guides provided to the students left a possible interpretation as was sought. The guidance was not total.

For well-rated students, it is difficult to know if the introduction of functional analysis is beneficial to them, because they stay good students.

10.5 Example 2: The Systemic Approach in Technology Education: Effects of Transferred Learning in Physical Science Problem Solving

We talked about the gradual construction of learning from one level of education to the next. The study presented below examined whether students (15–16 years old) who have received technology education on a systemic approach of industrial systems are better than other pupils (of the same age but from other academic domains such as literary ones or ones that are economics-based) at solving physical science problems which involve systemic reasoning. It shows that beyond the tool, functional analysis is a philosophy, a reasoning, a way of thinking in its own right that is a positive addition to the natural reasoning of the students.

At this level technology education becomes optional. The option is called Introduction to Engineering Science (ISI). The curriculum clearly specifies functional analysis, structural analysis, and technical system behaviour analysis (BOEN, 2001) as a systemic approach. Therefore, we can hypothesize that pupils who learn to develop a systemic approach in the option are more able than other pupils (who do not learn the systemic approach) to solve all sorts of systems-based thinking problems (technical, biological, physical sciences, etc.).

Following previous studies on natural thinking processes in a hydrodynamics system (Blondin et al., 1992), which showed the difficulties that students encounter in applying systems thinking, we have duplicated the task previously studied by these authors and extended it to an electric circuit that presents similar difficulties. The types of reasoning used by pupils to describe those systems in areas of physical sciences are characterized as natural, local, sequential or constant flow reasoning (Closset, 1989; Joshua & Dupin, 1993; Tiberghien, 2003; Viennot, 2006).

10.5.1 Method

Data collection: Students have to solve two problems:

Problem one (P1) (Fig. 10.2) is the same as used by Blondin et al. (1992).

In this hydraulic system, the fluid circulates using a pump (P). There is a narrowing (R). In a first situation (S1), students must: compare the quantity of water that flows per second between A and B, then between B and C. In a second situation (S2), students should assess the flow at each point (A, B, C) and take into account the shrinkage.

The second problem (P2) applies the same questioning about an electrical circuit including a current generator and a resistor R (Fig. 10.3).

Firstly, students must compare electrical intensity upstream (point A) and downstream (point B) of the resistor. Then, students have to say whether the electrical intensity changes at each point (A, B) after an increase of the resistor (between A and B).

This kind of problem can be both solved without specific knowledge of physical science: a systemic view based on the "conservation principle" of fluid within a closed circuit, is sufficient to answer correctly.

Sample: We collected papers answers from 74 students in year 12 and answers from 41 students in year 13. Among students of year 12, 58 had previously studied

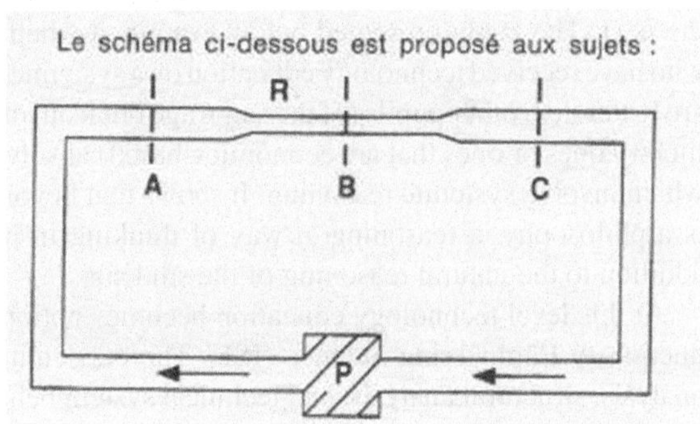

Fig. 10.2 Hydraulic circuit (P1)

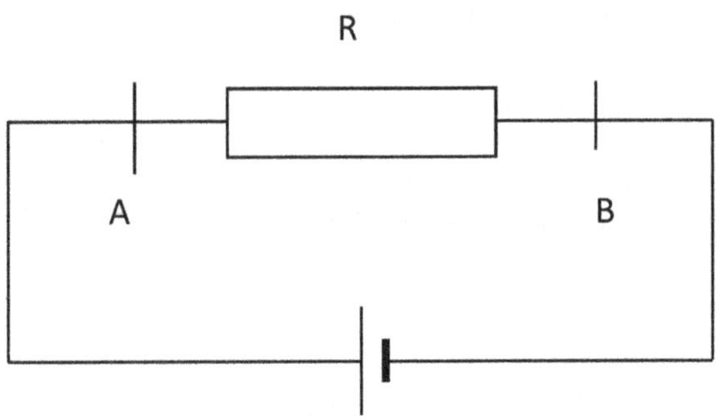

Fig. 10.3 Electric circuit (P2)

Table 10.7 % of students according to each pattern of responses observed on flow comparison at points A, B and C (Nota: correct pattern of answers is in bold)

Types of answers	Answers relating to flow at A, B, and C	Year 12		Year 13	
		With ISI	Without ISI	With ISI	Without ISI
1	**A = B = C**	**35**	**6**	**26**	**40**
2	A > B < C	17	56	16	20
3	A > B = C	24	12	16	20
4	A > B > C	7	0	13	0
5	A < B > C	17	26	19	0
6	A < B = C	0	0	7	0
7	A < B < C	0	0	3	20
Total of answers		100%	100%	100%	100%
N students		58	16	31	10

the ISI option and 16 students took another option (Economic and Social Science). Among students of year 13, 31 students studied ISI. Ten pupils who did not study ISI took the option scientific (S). In year 13, students who stayed in the ISI option should be the ones who perform better if the ability to analyse technical systems reinforces their systemic reasoning.

10.5.2 *Results*

Responses regarding the hydrodynamic problem (P1)

Table 10.7 presents the % of students according to each pattern of responses observed on flow comparison at points A, B and C in S1

We notice a clear difference in year 12 where 35% of ISI students opt for a constant flow versus 6% of students who had not studied ISI. On the other hand, students without ISI from year 13 have better understanding of the situation than their friends who studied ISI (40% compared to 26%). Knowing it is easier to teach with the principles of hydrodynamics (cf. the results obtained by Blondin et al. from an experimental group) explains the gap between year 13 with and without ISI. For years 12 with ISI, the learning transfer is easier because it concerns the same area of knowledge. On the contrary, for non-ISI students, it would constitute a trans-disciplinary transfer. If we add all local points of view, we notice a small decrease in the performance of ISI students between year 12 and year 13: judgments in favour of a constant flow regress from 35 to 26%. This positive influence of the ISI option seems to be confirmed whether the students continue the ISI option afterwards.

Table 10.8 presents the % of students according to each pattern of responses observed the flow at each point (A, B, C) after pointing the narrowing at point B (S2).

Table 10.8 % of response on the comparison of flows at each point before and after increasing the narrowing

Types of answers	Comparison with the flow measured at the same place before increasing the pipe narrowing			Year 12		Year 13	
	Point A	Point B	Point C	With ISI	Without ISI	With ISI	Without ISI
1	**–**	**–**	**–**	**18**	**0**	**23**	**0**
2	=	=	=	23	6	14	40
3	=	–	–	13	26	27	20
4	=	–	=	9	13	3	10
5	=	+	+	2	6	3	10
6	=	+	=	14	6	14	10
7	=	=	+	2	6	0	0
8	+	–	–	7	13	3	0
9	+	–	=	4	6	0	0
10	+	–	+	2	6	3	10
11	–	–	+	0	6	0	0
12	–	=	=	2	0	10	0
13	–	+	+	4	6	10	0
Total				100%	100%	100%	100%
N students				58	16	31	10

Nota "–; = ; + ", respectively, means that the flow is considered inferior, equal, or higher than it was before the increase in contraction

This distribution confirmed the positive influence of ISI. Correct answers (in bold in the table) come exclusively from students who studied the ISI option. It seems that, in itself, ISI instruction is helping to destabilize the usual spontaneous conceptions in promoting systemic types of reasoning. Because answers in favour of a generalized flow reduction can remain underpinned by a sequential approach, only considering the arguments that justify these judgments can allow us to be sure of their truly holistic nature.

Most of the arguments given (responses 3 to 6) by ISI students in year 12 and 13 confirm the generally sequential nature of these responses. In these students' minds, the change made to the circuit cannot have any repercussion flowing upstream. For them at A and at C, the pipe stays the same, so the flow does not change. This sequential reasoning is also evident in the answers given by those who think that the flow at B is greater than it was previously (answers 5 and 6). Some subjects affirm that the flow upstream from the pipe narrowing is greater than before (responses 8, 9, 10). We also notice that the dominant opinion (response 2) among year 12 ISI (23%), as in scientific year 13 (40%), corresponds to the natural way of thinking which consists of thinking about the fact that each circuit is characterized by a flow (or current)

Table 10.9 Distribution (%) of answer types about the comparison of the intensity at points A and B

Answers type	Intensity answer in A and B	Year 12 c		Year 13	
		With ISI	Without ISI	With ISI	Without ISI
1	**A = B**	**57**	**38**	**48**	**60**
2	A > B	41	62	48	40
3	A < B	2	0	4	0
Total		100%	100%	100%	100%
N students		58	16	31	10

value, regardless of the possible modifications to other parameters (differences in pressure or potential, resistances…).

When the holistic and systemic concept is present, it is the case of ISI, we observe that it induces positive changes in the results of student's ability to analyse technical system.

Responses regarding the electricity problem (P2)

The principle of the electrical circuit presented to students is the same as in the previous problem but now with a generator and a resistance included in the circuit.

Table 10.9 presents the % of answer types about the comparison of the intensity at points A and B (S1)

The majority of responses are divided on question 1 (good answer) in favour of maintaining intensity inside the electric circuit and question 2. The influence of knowledge learnt at school (question 1) is clearly highlighted by comments expressed by student in favour of maintaining intensity inside the electric circuit, like "*the intensity is identical at every point in a series circuit*" or by the formula "$V = RI$ ". It is also highlighted by arguments of question 2. Students (question 2) affirm that intensity is lower further down from resistance than further up. The natural tendency to adopt a sequential thinking is still very common.

The majority (62%) of year 12 non-ISI pupils affirm that current intensity is lower further down from resistance than further up because the resistance "*takes a little bit of current*", "*retains some of the electrons*", "*there isn't a bulb in the circuit*", or "*because A is situated next to the positive terminal*" or "*because the intensity at A is equal to that of the generator*". We see the same phenomenon with non-ISI pupils.

The proportion of students who succeed in each domain appears greater for ISI students who have studied technical systems.

Table 10.10 presents the % of types of answer on the comparison of intensity after an increase in loss of charge.

One-third of students says that the intensity is lower after the introduction of resistance into the circuit (answers type 1), but only 14% of them (non-ISI in year 12) gave this correct pattern of answers.

For most pupils, increasing the resistance only changes the intensity at point B (answer type 3) meaning downstream from the place where it is situated. This point

Table 10.10 % of answer on the comparison of intensity after an increase in loss of charge

Types of answers	Comparison with the intensity measured at the same point before resistor R increase		Year 12		Year 13	
	In A	In B	With ISI	Without ISI	With ISI	Without ISI
1	**–**	**–**	**29**	**14**	**32**	**30**
2	=	=	17	22	10	20
3	=	–	45	57	39	30
4	=	+	2	0	0	0
5	+	–	0	7	16	0
6	+	+	0	0	0	10
7	+	=	2	0	0	0
8	–	+	2	0	0	0
9	–	=	3	0	3	10
Total			100%	100%	100%	100%
N students			58	14	31	10

Nota "–; = ; + ", respectively, means that intensity is considered inferior, equal or higher than it was before the resistor increase

of view is slightly less common among year 13 pupils (39% et 30% compared to 45% and 57% for year 12). The numerous arguments to the location of R are clear with regard to the sequential nature of the reasoning used.

Correct answers at S1 and S2 are given by 26% of ISI and 14% of non-ISI pupils in year 12, and 32% of ISI and 20% of non-ISI pupils in year 13. Here, the constant intensity reasoning (response 2) is made by 10% of former ISI pupils in year 13, and 20% of pupils in the other groups.

For most pupils, increasing the resistance only changes the intensity at point B, meaning downstream from the place where it is situated. However, this point of view is slightly less common among year 13 pupils (39% compared to 45% for year 12 ex ISI and 30% versus 57% for year 12 non-ISI pupils). The numerous references to the location of R are clear regarding the sequential nature of the reasoning used.

It should be pointed out that the effect of the ISI option appears less clearly, when the problem is more familiar to all students, as it is in the case in this electric system. This study shows that ISI students who have learned about systems in a systemic and functional way are able to analyse all kinds of systems better than other students.

10.6 Conclusions

Our result on the first study shows that the teaching of functional analysis in technological education facilitates the transfer of a systemic reasoning to new situations and contributes to destabilizing the habitual naive approach of students. It is a plus for students who have this education. Functional analysis appears to be an effective means for both teachers and students. For the teacher, it is the instrument that allows both structure and transmission of knowledge. For the student, it will be an accessible systemic approach that provides specific skills for reinvestment.

Systemic thinking is a complex activity, and its teaching as well as learning is context specific. This work must be done in terms of technological education. At present, learning using a systemic method is really put into practice in higher education and in technical studies. While it is important that future designers of our technical systems consider all the dimensions of the objects around us, users, consumers and repairers will necessarily have to make decisions about functional interfaces associated with systems. Learning must be progressive, and there is a lack of data related to this which inhibits the effectiveness of learning.

Our results on the second study show that a functional approach focused on the understanding of structure and functioning of systems and promotes (with regard to the hydraulic circuit) an evolution of the conceptions with a reduction in "local" points of view, to the benefit of sequential reasoning. However, this positive impact only seems to be short term, unless this effect is enhanced by a continuation of this teaching (ISI) in years 12 and 13.

Concerning the generalized decrease in consecutive flow to an increased loss of charge within the circuit, the results of the second situation also seem to generate progress: the teaching of the technology option facilitates, for certain students, the transfer of a systemic reasoning to new situations, whereas for other students, it contributes to destabilizing the habitual naive approach.

The second problem (electro-kinetics problem) allowed us to note the unstable nature of the points of view adopted when students are faced with analogous circuits. A greater familiarity with the electricity domain than with the hydraulic domain explains the positive impact of the technology option. It is less obvious in the case of hydraulics because this topic is not taught in high school.

This research dealing with specific systems shows the relevance of a systemic approach. It is adaptable to any kind of system, and the students are able to use it even in other subjects. It is a complementary reasoning, a kind of deep learning in regard of the current complexity of our world (human, scientific and technological).

References

Aïm, R. (2013). *L'essentiel de la théorie des organisations 2013*. Gualino éditeur.

Andreucci, C., Chatoney, M., & Ginestié, J. (2010). The systemic approach in technology education: Effects of transferred learning in physical science problem solving. *International Journal of Technology and Design Education*. https://doi.org/10.1007/s10798-010-9148-y
Barak, M. (1990). Imparting basics in technology through an instructional system for computerised process control. *Research in Science and Technological Education, 8*(1), 53–67.
Barak, M. (2007). Problem-solving in technology education: The role of strategies, schemes and heuristics. In D. Barlex (Ed.), *Design and technology—For the next generation* (pp. 152–167). The Technology Enhancement Program and the Nuffield Design and Technology Project.
von Bertalanffy, L. (1968). *Organismic psychology and systems theory*. Clark University Press.
Blondin, C., Closset, J. L., & Lafontaine, D. (1992). Raisonnements naturels en hydrodynamique. *Revue Française De Pédagogie, 100*, 71–80.
Brown, J. S., Collins, A., & Duguid, P. (1989). Situated cognition and the culture of learning. *Educational Researcher, 18*(1), 32–42.
BOEN, Bulletin officiel du ministère de l'éducation nationale et du ministère de la recherché. (2001). *8*, HS du 31 aout. www.education.gouv.fr/bo/2001/hs2/seconde2.htm
Chatoney, M. (2003). *Construction du concept de matériau dans l'enseignement des « sciences et technologie » à l'école primaire : Perspectives curriculaires et didactiques (Thèse de doctorat)*. Université de Provence.
Chatoney, M. (2005). Organiser des activités de production à l'école primaire sélectionner des matériaux avec des élèves de 6 ans. *Aster, 41*(1), 139–158.
Chatoney, M. (2013). Etudier, concevoir, fabriquer et utiliser des artefacts techniques. HDR, AMU.
Closset, J. L. (1989). Les obstacles à l'apprentissage de l'électrocinétique. *Bulletin De L'union Des Physiciens, 716*, 931–950.
de Rosnay, J. (1977). *Le Macroscope: Vers une vision globale*. Éditions du Seuil.
de Vries, M. (2005). *Teaching about technology: An introduction to the philosophy of technology for non-philosophers*. Springer.
Deforge, Y. (1993). *De l'éducation technologique à la culture technique: Pour une maîtrise sociale de la technique*. ESF.
Dorst, K. (2006). Design problems and design paradoxes. *Design Issues, 22*(3), 4–17.
Dubois, S. & Gartiser, N. (2005). L'impact du concept de problème sur son processus de résolution. Application à la conception de systèmes techniques. *Actes du 6° congrès international de génie industriel*, 7–10 juin, Besançon.
Gille, B. (1978). *Histoire des techniques*. Gallimard.
Ginestié, J. (2000). *Contribution à la constitution de faits didactiques en éducation technologique : Curriculum vitae et publications (HDR)*. Université de Provence.
Graube, G., Dyrenfurth, M. J., & Theuerkauf, W. E. (2003). *Technology education: International concepts and perspectives*. Frankfurt am Main: Peter Lang Pub Incorporated.
Gunther, F. (2016). *Étude de l'efficacité des outils de l'analyse fonctionnelle dans l'enseignement et l'apprentissage de systèmes techniques au collège*. Aix Marseille Université.
Howard, R. A. (2007). The foundations of decision analysis revisited. In *Advances in decision analysis* (Vol. 1, pp. 32–56). Cambridge University Press.
Inhelder, B., & Cellerier, G. (1992). *Le Cheminement des découvertes de l'enfant: Recherche sur les microgenèses cognitives*. Delachaux et Niestlé.
Johsua, S., & Dupin, J.-J. (1993). *Introduction à la didactique des sciences et des mathématiques*. PUF.
Jouineau, C. (1982). *L'Analyse de la valeur: Méthodes, mise en oeuvre, applications*. Ed. Techniques Ingénieur.
Lebahar, J.-C. (2006). Pratique professionnelle et enseignement de la technique d'organigramme en architecture : Problèmes de transposition didactique.
Le Moigne, J. L. (1999). *La modélisation des systèmes complexes*. Dunod.
Martinand, J.-L. (2003). La question de la référence en didactique du curriculum. *Investigações Em Ensino De Ciências, 8*(2), 125–130.
Morin, E. (1990). *Introduction à la pensée complexe*. ESF Editeur.

Rabardel, P. (1995). *Les hommes et les technologies: Approche cognitive des instruments contemporains* (Vol. 1–1). A. Colin.

Scaravetti, D. (2004). *Formalisation préalable d'un problème de conception pour l'aide à la décision en conception préliminaire*. ENSAM

Simon, H. A. (1996). In *Models of my life* (1st MIT Press ed). MIT Press.

Simondon, G. (2012). *Du mode d'existence des objets techniques* (Vol. 1–1). Aubier.

Tiberghien, A. (2003). Des connaissances naïves au savoir scientifique. In M. Kail & M. Fayol (Eds.), *Les sciences cognitives et l'école. La question des apprentissages* (pp. 333–443). Presses Universitaires de France.

Vernat, Y. (2004). *Formalisation et qualification de modèles par contraintes en conception préliminaire*. Arts et Métiers: ParisTech.

Viennot, L. (2006). Modélisation dimensionnellement réductrice et traitement particulaire dans l'enseignement de la physique. *Didaskalia, 28*, 6–32.

Zehtaban, L., & Roller, D. (2012). Systematic functional analysis methods for design retrieval and documentation. *International Journal of Computer, Electrical, Automation, Control and Information Engineering, 6*(12), 1015–1020.

Marjolaine Chatoney is Professor of Education in the Institut National Supérieur du Professorat et de l'Education in Aix-Marseille University, France. She is member of the Research Unit 4671—ADEF (Learning, Didactics, Evaluation, Training) in which she directs the research program EAST (Efficiency of Artifacts in Science and Technology). She supervises research students in STEM, Design and Technology Education. Her research focus is on the role of artifacts, instruments and graphical intermediaries in processes of teaching and learning about design activities, robotics, and more broadly activity about technological systems. Her research contributions include: identifying the references in TE and epistemology; Reasoning and understanding systems; Gender-related phenomena in scientific and technological education; Education and access to knowledge for all (boys and girls from primary to high school); numerous scientific publications, supervision of theses and direction of research, editing and participation in international projects, member of scientific committees.

Fabrice Gunther has a PhD in educational sciences and a master in IT, and he is qualified as a lecturer. After working several years in the aeronautical and space industry and then teaching technology in different middle schools in France, he now teaches STEM in a French American school in Seattle, United States. His research focuses on the use and the efficiency of specific methods to analyse technical systems in a systemic way including both learning and teaching processes. His main publications are: Gunther, F. (2016). Study of the efficiency of tools from functional analysis in teaching and learning technical system in middle school, PhD thesis, Aix-Marseille University. Didactic of the conception - Chapter 8: The tools from functional analysis: Artefacts to understand technical systems (2020) Co-edition UTBM – HEP Vaud.

Chapter 11
Investigating School Students' Knowledge About Technological Systems: Towards "Qualities of Knowledge"

Jonas Hallström, Claes Klasander, and Ann Zetterqvist

Abstract Technological systems as a school curriculum component is complex, under-developed and under-researched. In this chapter, we present results from an investigation of school students' knowledge about technological systems, hypothesizing the occurrence of different "qualities of knowledge". A test instrument was distributed to 26 groups of students (n = 56) in a Swedish grade eight class (14–15 year olds), and data analysis was carried out using a qualitative, hermeneutic method. The findings show that the students' qualities of knowledge regarding the overall structure of the systems was quite advanced, but the systems or the societal context were not elaborated upon with any detail. The purpose of the system could be connected to humans and society, but students did not offer a description of the overall purpose. The flows that the students described were only of matter (water, wastewater) but not energy or information. The system boundary was also elusive, except for waste coming out of the sewer system and other environmental consequences. Thus, the test instrument was useful to gauge students' qualities of knowledge, especially regarding system structure, but with respect to some other system aspects the validity might need to be further improved.

Keywords Technology education · Technological systems · Secondary school · Qualities of knowledge

---

J. Hallström (✉) · C. Klasander
Department of Behavioural Sciences and Learning, Linköping University, S-60174 Norrköping, Sweden
e-mail: jonas.hallstrom@liu.se

C. Klasander
e-mail: claes.klasander@liu.se

A. Zetterqvist
Department of Pedagogical, Curricular and Professional Studies, University of Gothenburg, S-40530 Gothenburg, Sweden
e-mail: ann.zetterqvist@ped.gu.se

J. Hallström and P. J. Williams (eds.), *Teaching and Learning about Technological Systems*, Contemporary Issues in Technology Education,
https://doi.org/10.1007/978-981-16-7719-9_11

11.1 Introduction

Teaching about technological systems is nowadays mandated in many school curricula and standards around the world (e.g., ACARA, 2017; ITEA, 2007), but it is still a challenge for technology teachers. Previous research in Sweden shows that the learning demands on students from some teachers was low and not at the level required by the curriculum, which presumably has to do with an insufficient understanding of systems even among teachers (e.g., Schooner et al., 2018). This may stem not only from the inherent complexity of technological knowledge (Mitcham, 1994; Vermaas et al., 2011) but also from the short history of technology education and the less developed teaching practices. Furthermore, technological systems as a curriculum component is complex, under-developed and under-researched. Even though the previous research is rather limited, a few conclusions about students' and teachers' understanding of technological systems can be drawn. Firstly, both students and teachers are better at understanding structure, input and output of a system than its behaviour. Control mechanisms, feedback and flows of information are particularly difficult to grasp. Secondly, students understand systems better when they are scaffolded by, e.g., teachers, and thirdly, students develop a deeper, more complex understanding of systems as they grow older, especially regarding the included components. Finally, the role of humans in and around technological systems is difficult to understand, probably because humans fulfil so many different roles as designers, users and operators and thereby function as crucial but multi-faceted components of the systems (e.g., Hallström & Klasander, 2017; Koski & de Vries, 2013; Mioduser et al., 1996; Örtnäs, 2007; Svensson, 2011).

However, there is as yet no agreed upon template for what constitutes qualitatively different knowledge of technological systems, although Svensson et al. (2012) did formative research towards this end. In this chapter, we present results from an investigation of school students' knowledge about technological systems, hypothesizing the occurrence of different "qualities of knowledge". The investigation utilizes empirical research data from an exploratory Swedish school development project aiming at better understanding different aspects of students' knowledge of technological systems, through a broad array of examples of systems.

Swedish technology education has shifted quite significantly in emphasis, content focus and teaching methods over the last four decades. The subject evolved from being a male-dominated optional school subject in lower secondary school, via a compulsory component in a more interdisciplinary subject area where technological aspects were linked to natural sciences, to a redirection in 1994 when the school subject technology got its own national curriculum, from year 1 to year 9. This also meant that the sociotechnical dimensions—the relations between technology, society and nature—were strengthened and that studies about technological systems got a special component in the curriculum that is still central (Hallström, 2009). However, even if the investigation is set in a Swedish national context, we argue that the findings are relevant to discuss in relation to an international technology education audience and context.

11.2 Definitions and Methodology

When learning about technological systems, some aspects are considered by students to be more difficult than others. For example, visible components in a system are easier to understand than abstract, invisible phenomena such as feedback loops or system borders, and linear systems are better understood than nonlinear systems (e.g., Arbesman, 2017; Hallström & Klasander, 2020). In order to further explore how students understand technological systems, we constructed test instruments that built on hypothesized qualities of knowledge about technological systems. A technological system can be defined as a purposeful collection of components or sub-systems, connections and flows between them, and a system border that delimits the scope of the system. From this foundation, the test instruments were designed to incorporate a broad array of aspects on system knowledge, which was based on theories of technological systems and supposed to give students opportunities to reflect upon and express, e.g., their knowledge of the functions of certain technological systems, how they are delimited, what components they include, what kind of resources that flow through the systems, the systems' relation to their surroundings, the systems' historical change, interaction between components, how the systems are controlled, etc. (e.g., Bijker et al., 2012; Hughes, 1983, 2004; Ingelstam, 1996, 2012; Klasander, 2010; Vermaas et al., 2011).

Related to the above definition, we included altogether four different aspects of technological systems, and in these, we hypothesized that different qualities of knowledge could manifest:

1. *System boundary and relation to the surroundings:* every system has a system boundary, by which it relates to a surrounding which can include other humans, nature etc.
2. *Purpose of the system*: for every system, one can identify one or several purposes.
3. *System structure and behaviour (modelling)*: the structure of a system can be described as made up of components with relationships between them.
4. *Resource flows in the system:* in every system, there are different flows of energy, matter and information (cf. Hallström & Klasander, 2017; Schooner et al., 2018; Svensson et al., 2012).

The above four aspects of technological systems knowledge can be regarded as building blocks in a solid understanding of technological systems, and in this sense, they are not hierarchical but rather complementary.

Within each aspect, however, there is an increased complexity of knowledge, of which some but not all elements may be needed for a higher order of understanding of technological systems. We thus define "qualities of knowledge" as stages or phases of an increased complexity of systems knowledge (cf. de Jong & Ferguson-Hessler, 1996; Friege & Lind, 2006). This way of thinking about increased knowledge depth or complexity shares certain similarities with the structure of the observed learning outcome (SOLO) taxonomy. It also describes knowledge qualities in the sense of increased complexity and focus on quality, from unstructured to generalizable knowledge—unistructural, multistructural, relational and extended abstract (e.g., Biggs & Collis, 1989; cf. also Wilson, 2009).

The research reported in this chapter used test instruments that contained some 20 different assignments, between two and five sub-questions each. The assignments could be focusing on, e.g., water supply and sewerage, the national electric grid, cars and road transport, smartphones, elevators, GPS or electric ovens. The test instruments were varied regarding different types of technological systems and distribution of present-day and historical examples. In this regard, we included different "starting points" for the examples of systems in the test instruments, in order to test various qualities of knowledge about technological systems. One of the starting points is the interface between the supposed system and the human beings using it, for example, a toilet, a smartphone or an ATM machine. Another starting point is a fairly well-known technological system, e.g., the railway system, and then, one moves from that wholeness and successively identifies important sub-systems and components. A third starting point is following the historical change—forwards from a prior point in time, or backwards from now—of a well-known and agreed upon technological system (Hallström & Klasander, 2020).

For this investigation, one particular test instrument about water supply and, by extension, sewer systems was employed (see Fig. 11.1, the House). The test instrument, the House, was arranged with four sub-questions (a-d) supposed to give the students opportunities to express answers showing different technological systems aspects (1–4) for the freshwater and sewerage systems centred on the house. Table 11.1 outlines what kinds of aspects (1–4) were meant to be gauged in the four sub-questions (a–d) of the test instrument.

Fig. 11.1 Image of the house, used in the test instrument to visualize and scaffold the students when answering the sub-questions

Table 11.1 Outline of the focus of the sub-questions (a–d) of the student test instrument about water supply and sewerage systems, in relation to the aspects of technological systems knowledge (1–4)

The sub-questions in the test instrument: the House	Built-in aspects of technological systems knowledge in the test instrument
(a) What is the freshwater system for? What important needs of the inhabitants of the house can the system fulfil?	Purpose of the system (2)
(b) What is the freshwater system for? What important needs of society can the system fulfil?	Purpose of the system (2)
(c) How does the freshwater system work? Start with drawing a simple image of the freshwater system. Exemplify with some parts of the system that you think are important and describe how these work together	System boundary and relation to the surroundings (1); Purpose of the system (2); System structure and behaviour (modelling) (3); Resource flows in the system (energy, matter, information) (4)
(d) Does the freshwater system have an impact on the environment? If so, how?	System boundary and relation to the surroundings (1)

11.3 Research Methods

To try out the test instrument with school students, we contacted teachers connected to the CETIS (Swedish National Centre for School Technology Education) national network of technology educators, in order to see if they were interested in participating in this project. Through word of mouth, we got a positive response from two secondary schools in a mid-sized Swedish city, so the sampling method used was so-called snow-ball sampling (Robson & McCartan, 2016). The test instrument about technological systems was subsequently distributed in two mixed-gender grade eight classes (14–15 year olds) consisting of 26 groups of students ($n = 56$). Altogether 26 copies of the test instrument "the House" were thus filled in by the students in these secondary school classes. The students worked in pairs or in groups of three selected by the teachers. The teachers made sure that the students could work for as long as they needed to be able to finish filling in the assignment/the test instrument; this took approximately 45 to 60 min.

A hermeneutic, qualitative method of analysis was employed when coding and categorizing the data, that is, single student texts and images were related to the whole body of texts and images, the four systems aspects, and the systems context in a reciprocal, re-interpretive way (Ödman, 2007; Robson & McCartan, 2016). Since a part of the data was made up of the students' drawings of components and systems, often with textual comments, we thus employed so-called iconotextual analysis—hermeneutic analysis of text and image in combination—in the analytical process (cf. Axell, 2015; Ihde, 1990). The analysis of the student data was performed based on the Swedish transcripts. Excerpts that are included as particularly illustrative examples were translated into English by the authors.

The hermeneutic-qualitative analysis was performed in stages, where each stage of the analysis led to a deeper understanding of the data. First, an initial reading of the whole data set was performed, whereafter further readings were carried out in conjunction with initial coding and subsequent categorization. Thereafter, student answers were further scrutinized in relation to specific aspects of technological systems knowledge (1–4) (see above and Table 11.1). The student answers were subsequently evaluated in terms of their qualities of knowledge under each aspect, focusing on the *differences* between increasingly more complex qualities. At this stage, we also investigated the potential of the SOLO taxonomy to inform the analysis of the students' answers in relation to qualities of knowledge. The presented findings constitute the collective picture of the students' knowledge of systems aspects of the water supply and sewerage systems.

The validity of the study was ensured by building the instrument and the analysis on previous studies (e.g., Svensson et al., 2012), and by carefully trying out the questions in several stages, with several actors contributing to the validity at each stage. Our results can only be seen as representative of the 26 groups of students, although we hope to generate inter-subjective validity through them (Larsson, 2005).

Throughout the research process, the ethical principles for research were followed. The teachers and participating students consented to participation, after having been duly informed about the investigation in line with ethical guidelines (Swedish Research Council, 2017).

11.4 Student Qualities of Knowledge

Our findings are here presented with respect to students' potential qualities of knowledge, in relation to increased complexity from unstructured to more relational and generalizable knowledge—in line with the SOLO taxonomy.

1. **System boundary and relation to the surroundings**

Questions in the test instrument involving the aspect of system boundary and relation to the surroundings offer possibilities not only to clarify the system as such, but also to relate it to the surroundings and other systems with increased complexity as the system is described not as isolated but as both having a kind of boundary and relation to other systems, both natural and technical.

Most student answers did not show a demarcation or imply a system boundary. One example of this is shown in Fig. 11.2.

There were answers which defined a system border, but at the same time, they also misconstrued the connection between the water system and the sewerage system by suggesting that the wastewater from the house goes to the wastewater plant and then directly back into the freshwater system. One example is presented below in Fig. 11.3:

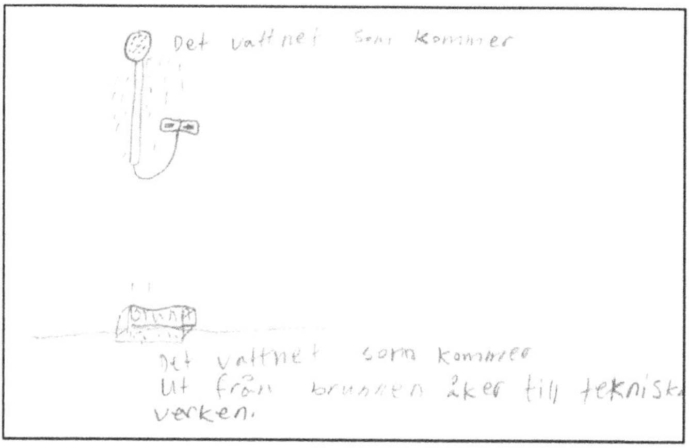

Fig. 11.2 Example of a system drawing without system boundary, showing a shower nozzle and a bathroom floor drain with accompanying text: "The water coming in […] The water that is discharged from the floor drain goes to *Tekniska verken* [municipal wastewater plant]."

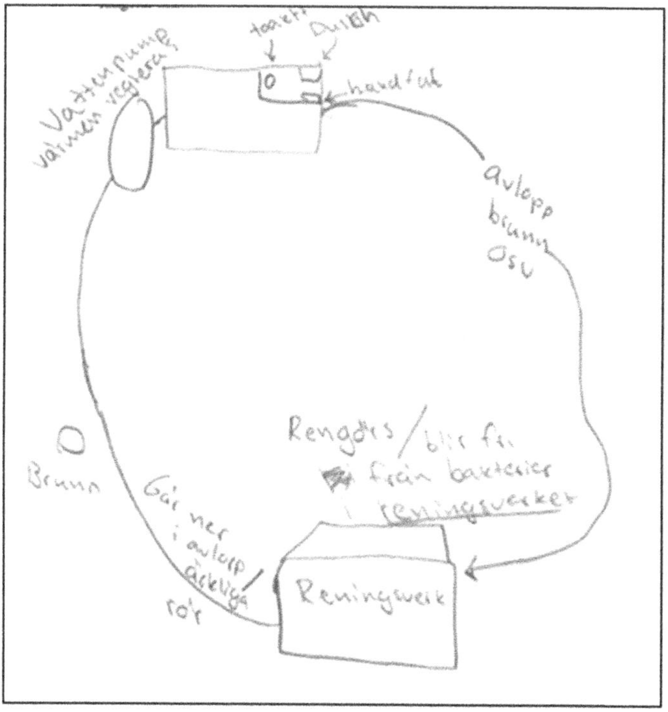

Fig. 11.3 Example of a system drawing of a closed loop between the water and sewerage systems. There are examples of miscellaneous components such as shower, sink, toilet, street drain, wastewater plant and pump

In reality, the purified wastewater is discharged into a nearby waterway and has to be purified again in a freshwater plant before it is fed back into the water supply system.

Most answers did not relate the system to its surroundings at all (see e.g. Fig. 11.2), thus showing low complexity. A few answers, on the other hand, related the system to other "systems", such as society and nature, showing both multistructural and relational features. Of these, most answers described negative influence in the form of various kinds of pollution:

> Yes, it affects nature in that particles are dissolved in the water [and] are discharged into the environment. Fish can get sterile from contraceptive pills. When farmers fertilize vegetation grows. When we build pipes we destroy a bit of nature.

There were also a few groups who included positive effects, for example:

> Water is discharged into the lake when it is clean, and then the animals can drink from the lake.

In summary, the student answers concerning this first aspect thus varied a great deal in complexity, with a predominance of less complex answers that did not recognize any system boundaries at all. However, some answers were more complex in recognizing system boundaries and even other technological and natural systems, although the relations described were mostly negative, for instance, various kinds of pollution in the environment.

2. **Purpose of the system**

When it comes to describing the purpose of the water system, we wondered if and how students described this in respect to the individuals in the house and to society. The questions addressing this aspect offer different possibilities to show the complexity in the purposes. For example, understanding the purpose for society involves more complex knowledge than the purpose for individuals.

Almost every student answer included that the purpose was to *provide* water to houses and society (fire brigades, swimming pools, factories, etc.):

> It is there to provide the residents with potable water and water for other needs. Washing, doing dishes, toileting, drinking, cooking, heating.
>
> The water system is there to provide clean water, to the drain, to fire hydrants to be able to put out fires. To swim in swimming pools. Sprinklers. To be able to make ice to ice rinks, for example. To hospitals/residential areas/apartments etc.

An observation from the analysis of the students' answers is that they were focused on giving *examples* on what the incoming water should be used for. All answers mentioned washing, almost every answer mentioned food preparation, but relatively fewer included heating. On a societal level, the examples were broader and fairly equally distributed between water used for fire brigades, hospitals, swimming pools, production companies, shops, heating, refuse collection and cleansing, water towers, schools and flooding. In summary, most answers focused on the input of water to

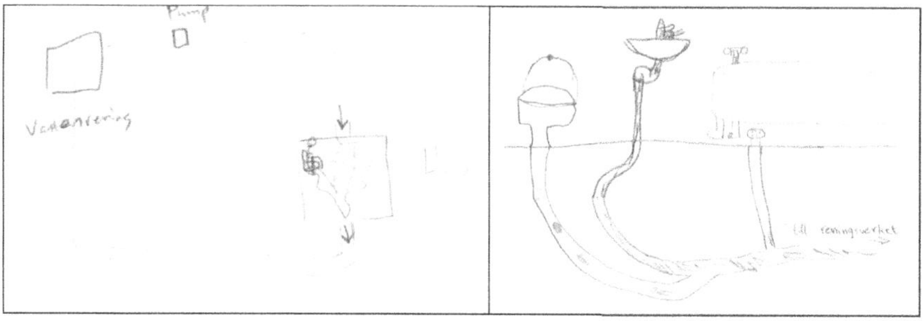

Fig. 11.4 Drawings of water systems with a few components; house, water treatment and pump in the left drawing by one group, and "to the water treatment plant" in the right by another group

houses and cities, rather than output. The examples given were richer on the societal level.

A minority of the answers described a purpose in terms of *leading away* water from houses; even fewer answers brought up discharge from parts of the city such as drains, wastewater and sewerage. Some indications on such purposes could, however, be found in the students' drawings, although not explicitly (see, for instance, Figs. 11.3 and 11.4). Some written examples of the purpose of leading away superfluous water are also given below:

> For everyone to get water and get rid of feces in sewers.
>
> There are in drains, basins to remove excess water that is later reused in purified form.

In summary, the complexity of the answers was fairly low regarding the purpose(s) of the water system. In relation to the use of resources, in terms of matter, mostly *freshwater* was mentioned. A few answers, however, stated that one purpose of the water system is related to providing *energy*, thus showing some example of a more complex reasoning:

> The water system is for heating the house through radiators and district heating and maintaining hygiene with hot water…

3. **System structure and behaviour**

The question that relates to this aspect gives students possibilities to both model the technological system and explain how it functions. The increase in complexity entails explaining relations between components in the system and how their behaviours influence different parts of the system as well as its purpose.

Regarding the most basic knowledge about a system's structure and behaviour, the majority of student answers exemplified components and specified relationships between components in the water and sewer systems. Two examples are shown in Fig. 11.4.

However, specifying how the components relate to the behaviour of the system was less well described by the students; here they were either vague about the relationship between components and the system as a whole, or they described a small part of

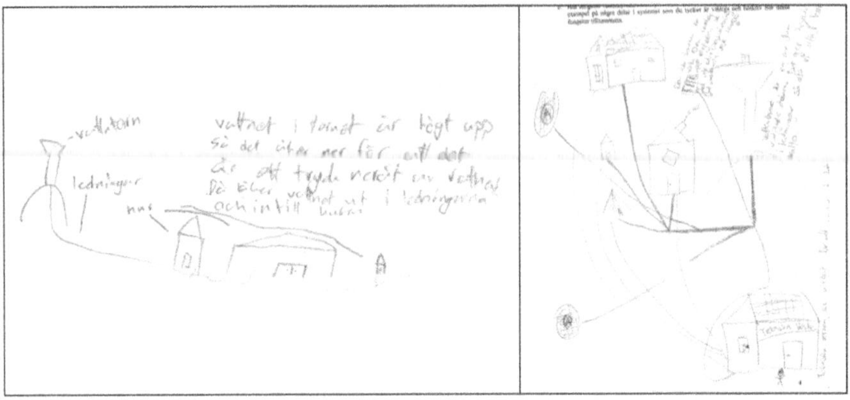

Fig. 11.5 Drawings of water systems with components relating to behaviour of the system. To the left one group writes: *"The water in the tower is high up so it goes down because there is a downward pressure of water. Then the water goes out into the pipes and into the house."* To the right another group writes *"…The water tower is one of the most important parts. It puts pressure in the pipes so we always have water…"*

the system (see e.g. Fig. 11.5). Fewer groups managed to show this quality than the previous, less complex ones.

The drawn models were of varying complexity, where a third of the answers included none or irrelevant models. Another third incorporated simple or incomplete models with few components (see Fig. 11.2), including several misconceptions. The remaining answers provided system models, of which one was a verbal model and the others visual (cf. Gilbert, 2004). Below is an example of a verbal model/analogy:

> A water system works like the blood system. […] The heart is like a water tower that keeps the pressure up all the time. The capillaries are like the system that pumps water.

Of the visual models, most were input–output models, but only a few of them also of a more complex network kind, such as Fig. 11.6.

This was also the only answer deemed to reach the more elaborate knowledge quality, to describe how changes to components/sub-systems influence other components as well as the purpose of the system. Thus, most of the models were basic such as of the input–output kind but lacked the more intricate descriptions of networks of components, and there were also some misconceptions such as in Fig. 11.7 below where purified wastewater feeds directly back into the freshwater system.

When summarizing the answers concerning this third aspect, we see a great variation. Most answers were of low complexity dealing only with components and relations between them. Of those that presented a drawing, few presented system models and only one of them reached a higher complexity in terms of the system's parts also influencing the purpose of the system.

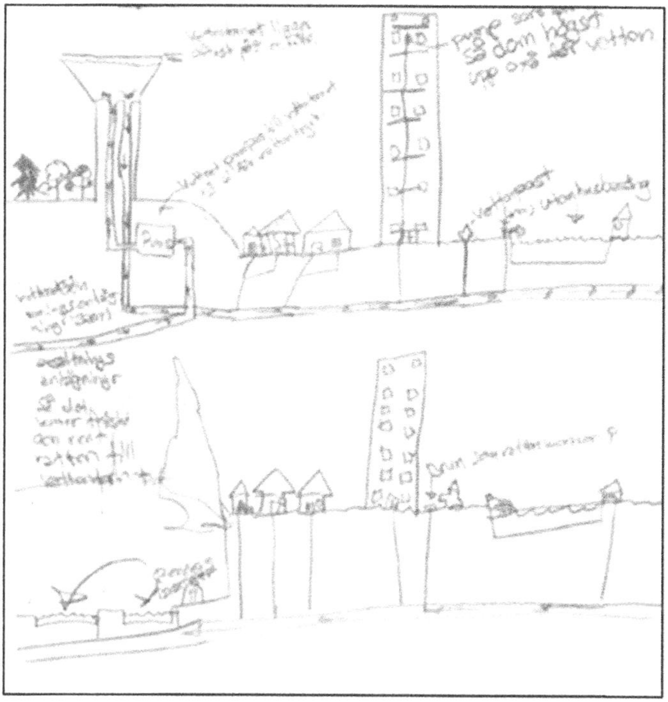

Fig. 11.6 Example of a system drawing showing the visible components as well as the pipe networks under-ground

4. **Resource flows in the system**

The aspect of resource flows is linked to the same problematic as the former aspect, to draw a model of the system and explain how the system functions. The increasing complexity, however, is about more abstract resource flows than matter, namely energy and information, for instance, how flows of information are used to control the system.

Not surprisingly, many students described flows of matter in the form of water. They described in words and/or pictures how the water flows from the freshwater plant and water tower through pipes to the house, and a few also depicted how wastewater is discharged from the house and goes to the wastewater plant. See Fig. 11.8 for two examples.

The students did not, however, describe flows of energy or information nor how these flows are used in the system, with one exception. One group correctly observed that hot water carries energy, although it is unclear if they mean hot water input into the house (in which case this really comes from district heating systems) or within the house (from e.g., a home heating system):

> The water system exists for us to be able to reuse the water again and again, and even be able to control the temperature. That is, get heating to the taps, etc.

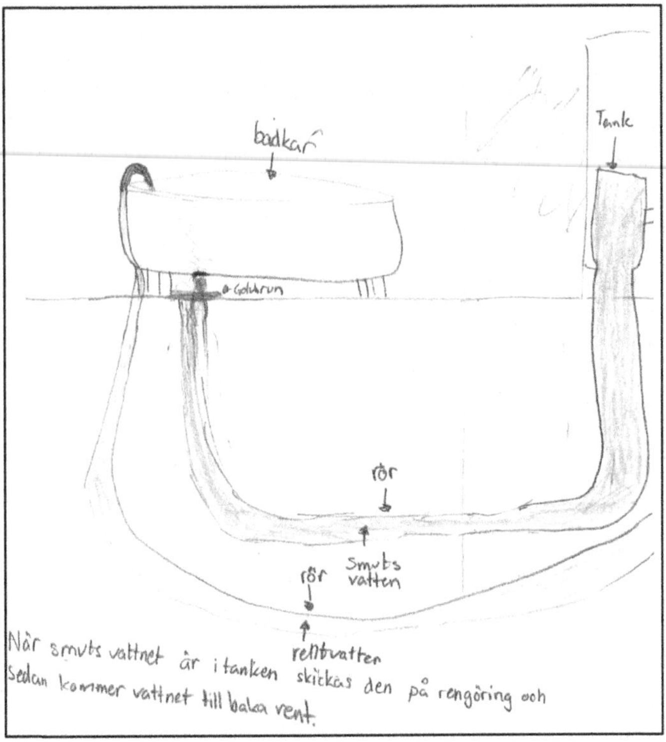

Fig. 11.7 Example of a system drawing of water and sewer pipes in the house where it is implied that purified household wastewater feeds directly back into the freshwater system

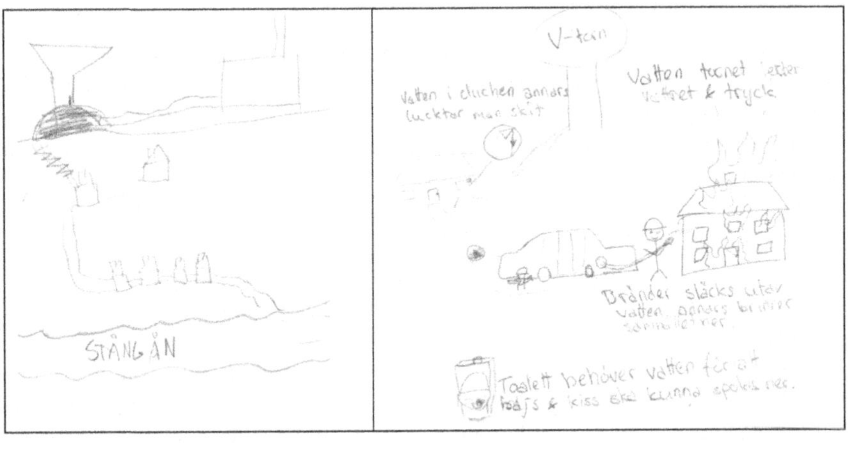

Fig. 11.8 Drawings of water systems only showing flows of matter (water). To the left is a drawing from one group, and to the right from another group that also writes: "*The water tower conducts water and pressure. Water in the shower otherwise you smell like shit. Fires are extinguished by water, otherwise, society burns down. Toilet needs water for poop and pee to be flushed down.*"

In summary, the answers about the fourth aspect, resource flows in the system, dealt only with concrete flows (water) and thus were of low complexity. None of the answers dealt with flows of information and only one answer mentioned energy, however not in terms of flows.

11.5 Summary

The findings indicate low levels of complexity related to the qualities of knowledge about a technological system. However, there were differences depending on the aspect. Regarding the **structure and behaviour** of the system, the answers were more complex than for other aspects, showing both knowledge of components and relationships between them, as well as some knowledge about how to describe the model and relating components to system behaviour. Regarding the **purpose** of the system, many examples were given about purpose both for humans and society but not related to resource flows or to other possible purposes. When it comes to the **system boundary and relation to the surrounding**, it was evident that boundaries seldom were described but some examples of relations to nature (often negative) were exemplified. No relationships with other systems were mentioned. The answers of the lowest complexity dealt with the aspect **resource flows** where the answers described flows of matter (water) but not flows of more abstract entities as energy and information. Neither were there descriptions of how energy is used within the system or how information is used to control the system.

11.6 Concluding Discussion

In this chapter, we present results from an investigation of school students' knowledge about technological systems, hypothesizing the occurrence of different "qualities of knowledge". The findings from the 26 groups of students reveal that the students have some knowledge of technological systems, but that it varies depending on the system in question and what aspects we are talking about. Especially regarding the aspects purpose and resource flows of a system, the students often did not show more complex qualities of knowledge, which we interpret is either because the teaching was not advanced enough or that our probes, the test items, did not allow them to express such understanding. For example, we only asked about "the water system" not "the water and sewer systems", while the figure in the test instrument shows arrows for both input and output (see Fig. 11.1) and the students sometimes referred to sewerage in their answers.

When the students answered, for example, the question about the purpose of the water system for the individual they often exemplified by mentioning activities (take a shower, drink water and flush the toilet), but they did not explicitly write about the purpose and main function of the water system. Our position when interpreting the student answers is thus parallel to the teachers' because we both try to identify if the student has shown his or her understanding of the purpose, etc., of the water system as a whole.

Consequently, the students showed quite advanced qualities of knowledge regarding the structure of the system, and some students could draw quite detailed systems near the house with freshwater plants and water towers. However, the systems within the home were not elaborated upon with any detail, nor was the wider societal context. The purpose of the system could be connected to humans and society, but students did not offer a definition of the overall purpose. The flows that the students described were only of matter/water, not energy or information, so there was no notion of the driving forces or control of the systems. The system boundary was also elusive, except for waste coming out of the sewer system and other environmental consequences. All in all, the reported investigation thus confirms much of previous research (e.g., Hallström & Klasander, 2017; Koski & de Vries, 2013; Mioduser et al., 1996). Thus, the only one of the four aspects of technological systems that seems to have "worked" is system structure and behaviour (#3), although students generally did not reach the most advanced qualities. Regarding the other three aspects (#1, 2 and 4), however, students never reached beyond the less advanced knowledge qualities (for a discussion of limitations of the investigation, see Hallström et al., 2021).

The description of depth or complexity of students' technological systems knowledge can also be conceptualized in line with the SOLO taxonomy. We therefore want to explore the possibility of relating the qualities of knowledge that we identified under each system aspect to the categories of the SOLO taxonomy. In Table 11.2, we show visually with colours where the most answers could be "located" in relation to four of the SOLO categories. The darker the colour, the more answers in the category. No colour at all means that none or only one single answer could be identified.

11.7 Implications

The reported investigation thus shows that school students' knowledge of technological systems seldom extends beyond the unistructural and multistructural SOLO levels. In actual fact, the teachers divulged that the participating students had not been taught about the water system in houses or society, and one student wrote: "This was the first time I heard the word 'water system'". In the textual answers and drawings, we could, however, see that the students had been educated about how a water tower works and how it distributes pressure in a system. We suppose that this might have been more a part of previous teaching in other subjects than technology. Students were also presumably taught about environmental pollution within science but not very much about the technological aspects of systems. Therefore, in order for the students to be able to develop more complex qualities of knowledge, they need to be taught more about technological systems (cf. Booth Sweeney & Sterman, 2007; Hallström & Klasander, 2017).

Table 11.2 Application of SOLO categories on four aspects of technological systems and the approximate abundance of student answers in different categories

	SOLO categories			
Aspect /quality	**Unistructural**	**Multistructural**	**Relational**	**Extended abstract**
1. System boundary and relation to the surrounding	No boundary described	Describe boundary	Relate to the system's surrounding: humans, society, nature and other systems (interdependence between systems)	Relate to several other systems and compare systems with similar purposes
2. Purpose of the system	Relate purpose to humans	Relate purpose to society	Relate purpose to use of resources (energy, matter, information)	Describe how questions about the purpose of the system can be answered on a systemic level (e.g. that it is possible to find several purposes of a system)
3. System structure and behaviour (modelling)	Exemplify components	Describe relationship between components	Relate components to system behaviour and describe the system using relevant model (e.g. network, or cyclical, model, hierarchical model, or input/output model)	Describe how changes to components/sub-systems influence other components as well as the purpose of the system
4. Resource flows in the system (energy, matter, information)	Describe flows of matter	Describe flows of energy and/or information	Describe energy that flows and is used in the system	Describe information that is used in the system for control purposes
	Concrete	**→ → → Increased level of abstraction → → →**		

In order to get students to understand relational and extended abstract aspects of systems, technology education needs to focus more on systems thinking and general features of systems, rather than on single, isolated examples. This goes for school classrooms and well as teacher education, and concerns particularly the first and fourth of our system aspects (see Table 11.2). Regarding the system boundary and the relation to the surrounding, students could be tasked to define various types of technological systems by identifying components, sub-systems and system boundaries. Teachers could help students visualize these structural aspects of the systems, and what components and sub-systems should be included depending on how the system boundary is drawn. It is important here to discuss the fact that, for example, a car could be a component in the road transport system, while the car is also in itself a system with many components and sub-systems. Furthermore, the resource flows of the system could also be seen as an important feature of the issue of delimitation, because it is often in the interface between the system and its surrounding that flows are exchanged, for example, matter in the form of environmental pollution that several students in our study could pinpoint. However, flows of energy and information are more complex to identify, especially as in particular the flow of information also contributes to the control of the system and thereby to its delimitation (Ropohl, 1979; Svensson et al., 2012).

As regards the purpose of the system, the arduousness of thinking in terms of the flows of the system may also be a reason why the students in our study had difficulty in understanding the aim(s) of the system. One way of enhancing students' understanding of a system's purpose is to have them discuss the more general purposes of technology, for instance, for *transformation, storage, transportation* and *control*, as these purposes have evolved historically (cf. Ropohl, 1979; van Wyk, 1984). Finally, we want to propose four distinct pedagogies for teaching about technological systems that address all of our four system aspects:

1. *Interface pedagogy*: Starting with the interface between the supposed system and the human beings using it. By starting, for instance, with the toilet seat, you move the students towards the other important components and the wholeness of either the sewer system or the freshwater system, depending on the direction.
2. *Holistic pedagogy*: Starting with a well-known technological system (e.g., the railway system), you move from that wholeness and successively identify important sub-systems and components, without succumbing to an overwhelming level of detail (so-called "black boxing").
3. *Historical pedagogy*: Following the historical change—forwards from a prior point in time, or backwards from now—of a well-known technological system, you can identify important structures, sub-systems and components, e.g., in the telephone system. With this method, it is also possible to identify some of the most common patterns of technological change (see chapter by Hallström and Kaijser).
4. *Design pedagogy*: All the above pedagogies are about analysing existing systems, but many curricula refer to the notion that technology education is about designing products or systems. Designing would include investigating, prototyping, and making working models of technological systems of appropriate complexity (Hallström & Klasander, 2020 p. 73–74).

References

ACARA, Australian Curriculum, Assessment and Reporting Authority. (2017). In *Australian curriculum, design and technologies*. Australian Curriculum, Assessment and Reporting Authority. https://www.australiancurriculum.edu.au/f-10-curriculum/technologies/design-and technologies/rationale/. Accessed 19 October 2020

Arbesman, S. (2017). *Overcomplicated: Technology at the limits of comprehension*. Portfolio.

Axell, C. (2015). *Barnlitteraturens tekniklandskap. En didaktisk vandring från Nils Holgersson till Pettson och Findus*. Ph.D. Thesis. Linköping University.

Bijker, W. E., Hughes, T. P., & Pinch, T. J. (Eds.). (2012). *The social construction of technological systems: New directions in the sociology and history of technology* (Anniversary Edition). The MIT Press.

Biggs, J., & Collis, K. (1989). Towards a model of school-based curriculum development and assessment using the SOLO taxonomy. *Australian Journal of Education, 33*(2), 151–163.

Booth Sweeney, L., & Sterman, J. D. (2007). Thinking about systems: Student and teacher conceptions of natural and social systems. *System Dynamics Review, 23*(2/3), 285–312.

De Jong, T., & Ferguson-Hessler, M. G. (1996). Types and qualities of knowledge. *Educational Psychologist, 31*(2), 105–113.
Friege, G., & Lind, G. (2006). Types and qualities of knowledge and their relations to problem solving in physics. *International Journal of Science and Mathematics Education, 4*(3), 437–465.
Gilbert, J. (2004). Models and modelling: Routes to more authentic science education. *International Journal of Science and Mathematics Education, 2*(2), 115–130.
Hallström, J. (2009). Systemteori och teknik. En introduktion till stora tekniska system. *Världens gång–teknikens utveckling. Om samspelet mellan teknik, människa och samhälle.* Studentlitteratur.
Hallström, J., & Klasander, C. (2017). Visible parts, invisible whole: Swedish technology student teachers' conceptions about technological systems. *International Journal of Technology and Design Education, 27*(3), 387–405.
Hallström, J. & Klasander, C. (2020). Making the invisible visible: Pedagogies related to teaching and learning about technological systems. In P. J. Williams & D. Barlex (Eds.), *Pedagogy for technology education in secondary schools—Research informed perspectives for classroom teachers.* Springer.
Hallström, J., Klasander, C., & Zetterqvist, A. (2021). Towards a student systems thinking inventory: Defining 'Qualities of Knowledge' about technological systems. *Techne Series: Research in Sloyd Education and Craft Science, 28*(2), 196–203.
Hughes, T. P. (1983). *Networks of power: Electrification in Western Society, 1880–1930.* Johns Hopkins University Press.
Hughes, T. P. (2004). *Human-built world: How to think about technology and culture.* University of Chicago Press.
Ihde, D. (1990). *Technology and the lifeworld: From garden to earth.* Indiana University Press.
Ingelstam, L. (Ed.) (1996). *Complex Technical Systems.* Swedish Council for Planning and Coordination of Research (FRN).
Ingelstam, L. (2012). *System—att tänka över samhälle och teknik* (2nd ed.). Energimyndigheten.
International Technology Education Association. (2007). *Standards for technological literacy: Content for the study of technology* (3rd ed.). International Technology Education Association.
Klasander, C. (2010). *Talet om tekniska system: Förväntningar, traditioner och skolverkligheter.* Ph.D. Thesis. Linköping University.
Koski, M.-I., & de Vries, M. J. (2013). An exploratory study on how primary pupils approach systems. *International Journal of Technology and Design Education, 23*(4), 835–848.
Larsson, S. (2005). Om kvalitet i kvalitativa studier. *Nordisk Pedagogik, 25*(1), 16–35.
Mioduser, D., Venezky, R. L., & Gong, B. (1996). Students' perceptions and designs of simple control systems. *Computers in Human Behavior, 12*(3), 363–388.
Mitcham, C. (1994). *Thinking through technology: The path between engineering and philosophy.* University of Chicago Press.
Robson, C., & McCartan, K. (2016). *Real world research* (4th ed.). John Wiley & Sons.
Ropohl, G. (1979). *Eine Systemtheorie der Technik.* Carl Hanser Verlag.
Schooner, P., Klasander, C., & Hallström, J. (2018). Swedish technology teachers' views on assessing student understandings of technological systems. *International Journal of Technology and Design Education, 28*(1), 169–188.
Svensson, M. (2011). *Att urskilja tekniska system. Didaktiska dimensioner i grundskolan.* Ph.D. Thesis. Linköping University.
Svensson, M., Zetterqvist, A., & Ingerman, Å. (2012). On young people's experience of systems in technology. *Design and Technology Education: An International Journal, 17*(1), 66–77.
Swedish Research Council. (2017). *Good Research Practice.* Stockholm.
van Wyk, R. J. (1984). Panoramic scanning and the technological environment. *Technovation, 2*(2), 101–120.
Vermaas, P., Kroes, P., van de Poel, I., Franssen, M., & Houkes, W. (2011). *A philosophy of technology: From technical artefacts to sociotechnical systems.* Morgan & Claypool Publishers.

Wilson, M. (2009). Measuring progressions: Assessment structures underlying a learning progression. *Journal of Research in Science Teaching, 46*(6), 716–730.
Ödman, P.-J. (2007). *Tolkning, förståelse, vetande. Hermeneutik i teori och praktik.* Norstedts.
Örtnäs, A. (2007). *Elevers vardagsuppfattningar om tekniska system. Bachelor's thesis.* Linköping University.

Jonas Hallström is Ph.D. and Professor of Technology Education at the Department of Behavioural Sciences and Learning, Linköping University, Sweden, where he is also currently Deputy Head of Department. He is also Chair of the Swedish National Graduate School in Science and Technology Education Research (FontD). He presents regularly at international conferences and consults on technology education. He is one of the editorial board members of the Brill/Sense book series *International Technology Education Studies*. His research primarily concerns the historical emergence of technology as knowledge content in the school, the epistemology and subject philosophy of technology, various subject content (e.g. technological systems) as well as the attitudes to and knowledge of technology and technology education of students, student teachers and teachers. The research also relates to technology teaching in relation to, e.g. design, gender (girls and technology), authentic learning, models and modelling and STEM (science, technology, engineering, mathematics) education.

Claes Klasander is PhD and since 2014 Director of CETIS, The Swedish National Centre for School Technology Education at Linköping University. His PhD thesis (2010) deals with how different actors in the Swedish school system have talked about teaching subject content relating to technological systems—rhetoric and praxis on three arenas. Since then, he has carried out several studies focusing on how students and/or teachers conceive of technological systems.

Ann Zetterqvist is PhD and Senior lecturer in Science Education at the Department of Pedagogical, Curricular and Professional Studies, University of Gothenburg, Sweden. Her current work includes PISA Science, teaching of subject matter didactics in science, and development of assessment instruments in science subjects. Her research interests include students' conceptions of technological systems and summative assessment of science knowledge.

Part IV
Conclusion

Chapter 12
Teaching and Learning About Technological Systems: A Research Synthesis

Jonas Hallström

Abstract Early twenty-first-century society is permeated by different kinds of technological systems. Such systems attract a great deal of research in different academic disciplines, but in education, research on technological systems as a comprehensive element in schools, tertiary institutions and universities is limited. This edited book was written by a group of scholars from a variety of educational disciplines with the majority from technology, engineering and science education, with the goal of summarizing, synthesizing and possibly expanding current research on the teaching and learning of technological systems. The aim of this final chapter is to synthesize research on the teaching and learning of technological systems as it has been presented in the book.

12.1 Introduction

Technological systems are essentially products of industrial society but as we enter a post-industrial era in the early twenty-first century such systems will only increase in importance. Ours is a networked and systemic society, being increasingly permeated by pipes, cables, servers, automation and signals. The expansion of an increasing variety of technological systems in society has also made them subject to research in more academic disciplines, not only STEM disciplines such as technology, engineering and science but also social science, logistics, economics, history, statistics and complexity science. In education, however, research on technological systems as a comprehensive element in schools, tertiary institutions and universities is limited. This edited book was written by a group of scholars from a variety of educational disciplines with the majority from technology, engineering and science education, with the goal of summarizing, synthesizing and possibly expanding current research on the teaching and learning of technological systems. The aim of this final chapter

J. Hallström (✉)
Department of Behavioural Sciences and Learning, Linköping University, S-60174 Norrköping, Sweden
e-mail: jonas.hallstrom@liu.se

J. Hallström and P. J. Williams (eds.), *Teaching and Learning about Technological Systems*, Contemporary Issues in Technology Education,
https://doi.org/10.1007/978-981-16-7719-9_12

is therefore to synthesize research on the teaching and learning of technological systems as it has been presented in the book.

The chapter will be divided according to the same sections that make up the book, namely I: Foundations of a technological systems philosophy; II: Contents, concepts, and contexts for teaching about technological systems; and III: Learning and teaching about technological systems.

12.2 Foundations of a Technological Systems Philosophy

The rationale for scientific inquiry regarding philosophical aspects of technological systems in the first part of the book is to investigate what technological systems *are*. Researchers dealing with this problematic have asked questions such as: What is a technological system compared to an artefact? How did/do technological systems evolve? What are the impacts of/on technological systems in relation to society and the environment? What are central technological systems concepts? The research underlying the philosophical aspects of technological systems was by and large conducted in the philosophy of technology and engineering, history of technology, sociology of technology, systems research, technology education and Science and Technology Studies (STS).

Built upon earlier systems research by, for example, Bertalanffy (1973) and Wiener (1948), historians and sociologists continued research in the 1980s with seminal studies by, for example, Hughes (1983, 1987) on the historical evolution of electric grids. Philosophers of technology later developed studies of engineering systems and the characteristics of global technological systems such as the civil aviation system (Kroes et al., 2006; Vermaas et al., 2011). Philosophers of technology such as Günter Ropohl also defined components of a general technology education, based on a conception of systems theory and technological systems (Ropohl, 1997, 2009). Researchers from different disciplines approached the study of technological systems differently, but all in all a philosophy of technological systems—what characterizes such systems, how they evolve, their relationship to society and the environment, their ubiquity, complexity and "impact", etc.—has a fairly well-established foundation.

Because technology education deals a lot with designing and making, it is by tradition focused on artefacts (e.g. Jones et al., 2013), so efforts to introduce systemic aspects of technology, such as the one by Ropohl, have only been influential in certain countries, for example, New Zealand and in some German-speaking countries (Compton, 2019; de Vries et al., 2020). There have thus been few concerted attempts at investigating the crucial characteristics of technological systems for technology and engineering education (exceptions include Hallström, 2009; Hallström & Klasander, 2020; Klasander, 2010), so in that sense there is still a need for studies into a technological systems philosophy specifically for education, in technology and other related educational disciplines. In the following, I will extract the most important contributions from this book.

In Chap. 1, "Socially Constructed and Society Shaping: Investigating Characteristics of Technological Systems for Technology Education", Jonas Hallström and Arne Kaijser focus on the historical evolution and geographical extension of technological systems, as a way of specifying certain characteristics that are crucial for understanding technological systems. It is important for students to learn such basic characteristics of technological systems because they differ in crucial respects compared to single technological artefacts. Hallström and Kaijser thus specify these characteristics in more detail, which includes technological systems being: socio-technical with both societal and technical components; developed by system builders and managed by professional organizations; with a spatial scope ranging from household to city-wide, regional, national and global networks; and dependent on control features including feedback loops as crucial mechanisms for making the systems stable. These technological systems also evolve—and sometimes devolve—in distinct phases and in particular societal, economic and geographical contexts, which may have repercussions when they are transferred to new contexts. Furthermore, the systems are dependent on each other which over time and space leads to an entanglement of systems. Technological systems also have a huge impact on the environment, which is why students will need to critically consider the human dependence on systems.

In Marc de Vries' chapter, "Technical systems and technical artefacts: how to conceptualize their relation" (Chap. 2), there is an ambition to clarify the meaning and use of concepts in technology education, in this case the relationship between a technological artefact and a system. This is important given the traditionally strong focus on artefacts in technology education. De Vries goes some way towards enhancing the clarity of the difference between a technological artefact and a technological system, but also, in a sense, of merging the artefact perspective and the socio-technical systems perspective, by conceptualizing systems with the philosopher Herman Dooyeweerd's ontology (Verkerk et al., 2016). De Vries connects both artefacts and systems to Dooyeweerd's taxonomy of reality, which consists of 15 aspects; the numerical (quantitative), spatial, kinematic (motion), physical, biotic (life), psychic (sensitive), analytical (logical), formative (historical), symbolic (linguistic), social, economic, aesthetic, juridical, ethical and belief (trust). He concludes that the social dimension of the most simple artefact (like a coin) and of the most complex system (like an airplane) should be taken into account by learners for an overall understanding of technology as a phenomenon.

Both the technical artefact and the socio-technical systems perspective could therefore be part of technology education. The extent to which the full complexity of technology in Dooyeweerd's fifteen aspects of reality is used, depends on the educational level. Dooyeweerd's aspects can come into play in relation to any artefact, but the more complex, systemic and "social" a technology is the more important the socio-technical systems perspective will be for understanding that technology. The difference between an artefact and a system is the greatest for the first of Dooyeweerd's aspects such as the numerical, spatial and biotic, because these aspects define the physical expansion and complexity (for example, number of components).

In conclusion for this section, although de Vries includes artefacts and systems under one conceptual umbrella, it is also quite clear from his Table 2 that there are differences between artefacts and systems, and an artefact perspective and a systems perspective. This is also obvious when we look at the previous research done in other fields, which is partially outlined above and is in fact also one of the rationales for developing this book; technological systems have certain salient characteristics that also require educational research, tools and approaches. Some of these characteristics were uncovered and outlined by Hallström and Kaijser, and some can be found in previous research, which together feeds into the next section of the book.

12.3 Contents, Concepts and Contexts for Teaching About Technological Systems

The rationale for scientific inquiry in the second section of the book is based on the philosophy section, but takes one step further to focus on investigating what are central curriculum perspectives on technological systems. Researchers dealing with this problematic have asked questions such as: What concepts of and perspectives on technological systems exist in technology curricula and standards? What is the contribution of systems thinking and systems theory? What are thus the central contents, concepts and contexts of technological systems appropriate for learning, in technology compared to adjacent fields (e.g. biology education)? What are some potential models for teaching about technological systems? The research forming the basis of this section was mostly carried out in technology education, engineering education, STEM education, systems theory and engineering disciplines.

There has been considerably less research on a technological systems curriculum, compared to the philosophical, sociological and historical research presented above, but there are still some previous studies worth mentioning. In terms of curriculum efforts based on research, Pearson and Young (2002) provided an early example of an integrated systems component. Compton and Compton (2013) also developed a curriculum framework for teaching about technological systems within the New Zealand curriculum which was based on extensive research on both students and teachers. Klasander (2010) went some way towards defining what technological systems contents, concepts and contexts are appropriate for technology education, based on analyses of systems theory, curricula, textbooks and teachers' conceptions (see also Klasander, 2006). The contribution of systems thinking/theory and modelling to a technological systems curriculum has been explored by Svensson (2018), and Hallström and Klasander (2020) also developed models and pedagogies for teaching about technological systems.

In Chap. 3, "Technological systems in national standards and curricula", Per Norström depicts aspects of technological systems and how they are included in national curricula and standards for technology education in primary and lower secondary school in Australia, New Zealand, South Africa, the USA and some European countries. He concludes that systems are seldom identified as a separate strand or theme but are supposed to be learned in conjunction with other technological phenomena. Furthermore, in curricula that emphasize designing and making, included systems are those that pupils can design and make from simple components and/or prefabricated sub-systems, whereas in curricula that include the social aspect of technology, infrastructure and other large systems tend to have more prominent positions. Cross-disciplinary systems thinking is mainly included in curricula where technology is combined with the natural sciences. The roles of non-technical components such as institutions or human agents in technological systems are not highlighted in the studied curricula, however.

In Chap. 4, "Cross-curriculum system concepts and models", Maria Svensson explores similarities and differences in the use of system concepts and models in technology education and biology education, to provide an opportunity to see recurring patterns in systems thinking between subjects and disciplines. Systems thinking could inspire or make up a pedagogical approach in which a holistic framework empowers both teachers and students to recognize how fundamental concepts taught in the classroom can be used as resourceful tools to better address complex, multi-component modern challenges (e.g. Ho, 2019). However, a unified conceptual framework for the development of an understanding of systems and systems thinking in education is still absent from most schools' curricula (Jacobson & Wilensky, 2006; Plate, 2010). Svensson thus presents and compares system concepts and models used in two different subjects, biology and technology, and reflects on how an awareness regarding system aspects in education may contribute to cross-curriculum learning opportunities. The chapter presents a qualitative literature review of identified concepts of the structure–behaviour–function (SBF) model in biology and technology education literature. Svensson concludes that the SBF model has potential as a pedagogical approach where structural and behavioural aspects of the two investigated subjects biology and technology can be compared. This could be the first step towards transfer of system knowledge and contribute to cross-curriculum learning opportunities (cf. Rosenkränzer et al., 2016).

In Chap. 5, "Fostering Systems Thinking in the Context of Electronics Studies", Moshe Barak also explores systems thinking, but in the context of electronics. Systems thinking is considered a major higher-order thinking skill essential for successful integration in areas such as science, technology, engineering, or management sciences. However, this term has remained somewhat vague and is defined or represented through diverse models or characteristics lists. The idea of developing systems thinking in the context of electrical and electronics systems stems from the fact that these are central areas in modern technology and are close to the students' world everywhere. In his chapter, instead of using the SBF model Barak identifies the following seven aspects of systems thinking specifically associated with electricity and electronics, specifically home room-heating devices, an air conditioner and a sound system: W—seeing the "Whole" rather than the "parts"; I—Interconnections/interrelationships; F—Feedback loops; D—Dynamic behaviour; B—System boundaries; T—Technological innovation; and S—Socio-technological implications. In the last section of the chapter, Barak also addresses four key aspects of teaching systems thinking in technology education: modelling; the STEM view; the engineer and the technologist; and innovation.

In Chap. 6, "An educational model for teaching about technological systems", Susanne Engström and Maria Svensson try out an educational model. In their chapter, they illustrate the process of taking the "Freiburg heuristic competence model of systems thinking" (Riess et al., 2015) which describes more general systems thinking capabilities, and transforming it to a model for teaching and learning about technological systems. Relevant knowledge and capabilities related to a common system example in technology education in Sweden, the wastewater system, are implemented within the structure of the model with the intention to investigate if the Freiburg model is useful and understandable as an educational model. Engström and Svensson show that the resultant model is largely helpful when transforming technological systems knowledge into actual activities for students in technology education. To become a useful educational model that infuses all aspects from the Freiburg model—structural, behavioural, functional, and contextual aspects of systems—into technological systems teaching, Engström and Svensson conclude that further development of the model is needed especially concerning contextual aspects. For students to develop an ability to discuss and analyse issues about sustainable development and solve environmental problems in society they need to have an understanding of contextual aspects (Klasander, 2010). Furthermore, the transferability of aspects when using the model for one system, such as the wastewater system, to another system, need to be investigated further. Therefore, the Freiburg model could be seen as a nascent educational model that has potential.

In conclusion for this section, the fact that the included systems in the curricula are not subsumed under a distinct heading or theme called "technological systems" probably testifies to its still unclear status as a curriculum component. The absence in the curricula of, for example, human agents in or around the systems is a striking omission, but it is also in accordance with previous research about how students conceive the management of systems (e.g. Hallström & Klasander, 2017). Two different models are presented for promoting systems thinking in teaching about technological systems, the SBF model and Barak's own W—I—F—D—B—T—S model. Their objectives differ somewhat, however, in that Svensson wants to promote transfer of system knowledge, contributing to cross-curriculum learning opportunities, while Barak seeks to foster systems thinking in the form of the capabilities included in his model. Engström and Svensson show in their chapter that their proposed educational model is helpful for transforming technological systems content into workable activities for students in technology education. To become a useful educational model that provides all four aspects from the Freiburg model to technological systems education—the structural, the behavioural, the functional, and the contextual—further development of the model is needed. The likeness to the SBF model is striking, although a possibly fruitful contextual aspect is also added here.

12.4 Learning and Teaching About Technological Systems

The basis for scientific inquiry in the third part of the book is the investigation of what students know, how they learn as well as how teaching and assessment could be designed, about technological systems. Researchers dealing with this problematic have asked questions such as: How do children and adults conceive of and learn about technological systems? How could teaching and assessment be formed about technological systems? The research contributing to these questions was largely drawn from technology education, engineering education, STEM education, engineering disciplines and learning theory.

There has been quite a lot of research in this area, especially during the last decade. However, as early as the mid-1990s Mioduser et al. (1996) investigated the mental models that American middle-school students have of control systems such as automatic doors, heating/cooling systems, and various household devices, before, during and after instruction. They concluded that the students' understanding and the "correctness" of their mental models prior to instruction were very poor, but a little better after. Control features as well as flow of information in a system were generally poorly understood, while system structure was better understood. Ginns et al. (2005) also carried out an intervention on technological systems in an Australian grade six class and noticed improvement in students' abilities to describe relationships between inputs, processes and outputs.

Koski and de Vries (2013) designed an intervention study in primary education and observed that the concept of input was clearer to the students than output, but that the latter conception improved somewhat after the intervention. Abstract phenomena such as flow of information in and boundaries around systems were challenging to conceive. Svensson's (2011) study of 10- and 15-year-old students' experience of technological systems concluded that the students understood the structure quite well but were not so knowledgeable about how components interact and how humans fit in the systems. Lind et al. (2019) similarly investigated 13–14-year-old students' understanding of technological systems and their characteristics, but in the context of studying the students' reasoning and collaboration in small groups. The researchers concluded that the extent of students' understanding was dependent on the context. Most of the students had no difficulty describing a technological system as consisting of different components and that these work together to create a desirable function, but socio-technical issues were increasingly difficult the more abstract and societal the systems became.

This is in line with results from Klasander (2010) on technology teachers, and he also concluded that systems thinking among Swedish technology teachers are often hampered either by a focus on scientific, reductionist aspects of systems or a focus on single artefacts. Svensson and Klasander (2012) also studied how two groups of technology teachers plan their teaching about technological systems in lower secondary school. The study shows that the teachers need more knowledge about the similarities and differences between various technological systems, as well as a better understanding of the system's components and different levels (physical, organizational, regulatory). Hallström and Klasander (2017) studied student teachers' conceptions of technological systems and concluded that the parts of the systems that the students could describe were mostly the visible parts such as components and devices, or the interface with a software. The "invisible" or abstract aspects, primarily flows of information or control operations, were more difficult to understand, however. Schooner et al. (2018a; b) also investigated technology teachers' conceptions about technological systems and found that while the teachers focused on the technological core of the system they could also express a socio-technical understanding where humans play a significant role.

In Chap. 7, "Young students' perceptions of control systems and adaptive robots", David Mioduser focuses on young students' development of a systems worldview while learning about control systems and adaptive robots. Young children encounter these artefacts in the form of, for example, sophisticated toys, computer-controlled games and devices, kitchen appliances, elevators and automatic doors, or traffic-lights systems. These kinds of devices, which began populating our world only a few decades ago, are characterized by purposeful functioning capabilities, autonomous decision-making, programmability, knowledge accumulation capabilities, and adaptive behaviour—challenging children's traditional and intuitive distinctions between the living and non-living realms, between inert and behaving agents, between the natural and the artificial. Based on evidence collected over many years by his research team, Mioduser elaborates on three main issues: (1) Young children's overall perception of adaptive systems; (2) Their detailed understanding of structural, functional and behavioural features of these systems; and (3) Implications of the research insights for supporting children's learning of adaptive systems concepts and skills.

Following the programming sessions, children developed complete, precise, complex and quite abstract mental models of the adaptive system—from spontaneous and undifferentiated models up to complete control models with differentiated control features and functions. In contrast to previous studies emphasizing kindergarten children's tendency to adopt animistic and psychological perspectives, Mioduser's team observed the early development of the use of technological (non-anthropomorphic) language as well as an engineering stance towards the adaptive systems. They observed a steady development in children's understanding of the structural (e.g. components, compounds), functional (e.g. flow of information) and behavioural (e.g. goal-oriented, emergent behaviours) levels of the systems. They also observed the development in children's capability to create appropriate representations of the programs required to control the system's adaptive behaviour, from linear event-like descriptions up to a-temporal rules.

Conversely, in Chap. 8, "Feedback in technological systems", Jonas Hallström reports that there was only a very slight improvement of adult secondary technology student teachers' conceptions of feedback in technological systems, after a two-cycle intervention with two consecutive student groups. The participating students were novices and the intervention consisted of a two-hour seminar and some preparatory reading about feedback in technological systems. Nonetheless, this could be considered a good start towards improving students' conceptions of feedback in technological systems, and it is encouraging that many students could develop a systemic understanding. Prior research on feedback in not just technological but also other kinds of systems showed that students of all ages find it very difficult to understand nonlinear systems in general, and feedback in particular (e.g. Arbesman, 2017; Hmelo-Silver & Azevedo, 2006). However, further studies need to be done to investigate how students' atomistic conceptions and device knowledge could be improved (Mioduser et al., 1996). Perhaps this could be done along the lines of the second cycle intervention of the study, with focus on creating block diagrams and designing simple technological systems with control mechanisms, although it needs to be more strictly planned, monitored and evaluated.

Chapter 9, "Student teachers' mental models of everyday adaptive control systems" by Osnat Dagan, also addresses student teachers' understanding of control systems. Use of adaptive control systems is constantly increasing, and more technological systems are becoming self-regulated. The understanding of how they work thus becomes more difficult as the processes are invisible, like a black box, which is why many people of all ages lack correct mental models of these systems. The aim of Dagan's chapter is to identify from prior studies and from her current research data what student teachers' prior adaptive control systems mental models are, and to design an instructional unit that will bridge the gap between these mental models and accurate ones. Undergraduate kindergarten and primary school student teachers at Beit Berl College (BBC), Israel, attended the "Technological thinking and robotics" course in which they learned about everyday adaptive control systems. It became clear that after the course students still did not have correct mental models of such systems that could assist them in life and as teachers. Therefore, the prior mental models of student teachers were examined. They were asked to describe how an automatic door works, either in sketches or in writing. Based on the findings arising from the current data, which reinforced earlier research findings (Hallström & Klasander, 2017; Koski & De Vries, 2013; Slangen et al., 2011; Svensson & Klasander, 2012), an instructional unit containing four activities was developed. These activities' main aim was to make the invisible visible, in order to clarify how self-regulation control systems work and help student teachers construct accurate "runnable" (Mioduser et al., 1996) mental models.

Chapter 10, "Learning and teaching with a systemic approach: using functional analysis in technology education for all", was written by Marjolaine Chatoney and Fabrice Gunther. The purpose of their chapter is to show that learning about technological systems, with tools designed and invented by engineers, also allows students and teachers to better understand the multi-technological and environmental complexity affecting the technological choices of the object or system concerned. The systemic approach makes it possible to consider the complexity of systems, and to do that engineers use tools that enable the study of a system function by function. This method of reasoning is called a functional approach, and the analytical tool associated with this is called functional analysis. Thus, the functional approach favoured in using these tools provides engineers with a systemic and complete analysis of a system. That is also why it is a fundamental part of the technology curriculum in France.

In their chapter, Chatoney and Gunther present results from two separate studies. The first one describes the uses of functional analysis in practices reported by technology teachers in middle school. It also reports on what students learn and can do with their learning when they are taught techniques from functional analysis. The second study measures the effect of transferred technical systems learning in physical science problem-solving in high school. It shows that students who have learned the systemic approach with the tool of functional analysis, develop a systemic reasoning benefit which is useful to understand other technological systems.

In Chap. 11, "Investigating School Students' Knowledge about Technological Systems: Towards 'Qualities of Knowledge'", Jonas Hallström, Claes Klasander and Ann Zetterqvist present results from an investigation of school students' knowledge about technological systems, hypothesizing the occurrence of different "qualities of knowledge". The authors designed a test instrument that was distributed to 26 groups of students ($n = 56$) in two Swedish grade eight classes (14–15 year olds), and data analysis was carried out using a qualitative, hermeneutic method. The findings show that the students' qualities of knowledge regarding the overall structure of the systems were quite advanced, but the systems or the societal context were not elaborated upon with any detail. The purpose of the system could be connected to humans and society, but students did not offer a description of the overall purpose. The flows that the students described were only of matter (water, wastewater) but not energy or information. The system boundary was also elusive, except for waste coming out of the sewer system and other environmental consequences. Thus, the test instrument was useful to gauge students' qualities of knowledge, especially regarding system structure, but with respect to some other system aspects the validity might need to be further improved.

In conclusion for this section, the research about students and student teachers' conceptions of technological systems presented in the book reinforce some findings from previous research: both children and adults understand quite well the structure of the systems, particularly "visible", concrete components, and flows of matter in the systems. Conversely, they generally do not understand black-boxed, "invisible" control features and flows of information as well, neither do they have deeper knowledge of the details of a system or how components work ("device knowledge"). When it comes to interventions to improve students' understanding of control features such as feedback, the findings in this section are inconclusive. Concerning technology student teachers, Hallström's intervention led to a slight improvement, whereas the results of Dagan's intervention remain to be seen. Findings from Mioduser and his team's research over the years, on the other hand, point to several important improvements in young children's mental models of adaptive control systems.

12.5 Conclusions, Implications and Instructional Approaches for Education About Technological Systems

In this chapter, I have summarized and synthesized prior research and research presented in this book, about the teaching and learning about technological systems. It is clear from the section conclusions of the chapter that the findings presented in the book as a whole both reinforce earlier research results and add some new findings. In this last section, I will draw some general conclusions about the learning of technological systems as well as suggest some possible fruitful instructional approaches adopted specifically to technological systems.

Learning about technological systems is complex because technological systems are in themselves complex, so there is no one clear-cut, patented way of learning. Instead, the chapters of the book suggest some tools for learning and some contexts in which learning can be effective. One obvious way to start is by *simplification*. De Vries suggests in his chapter a philosophical take on simplification. He claims that philosophy can bring out the essence of things and get away from all the details that make things complicated or map out the complexity to make it more comprehensible (as in the Dooyeweerd approach). Thus, in de Vries' view simplification could be both the reduction of detail, and the mapping out of complexity but in such a way as to make it more comprehensible (cf. the concept of complexification, Klasander, 2010). In this way, one does not confront learners with the full complexity of reality immediately, and a good technological systems philosophy could help us suggest what to start with and what to bring in at a later stage.

As indicated by Hallström and Klasander (2020, p. 79), in a classroom situation teachers could simplify by choosing examples of technological systems along the following lines:

- From simple to complex systems,
- From small to large and widespread systems,
- From systems related to myself via us to others,
- From local systems via regional/national to global systems.

All these four strategies involve a simplification, but it is not always obvious if this entails starting with the details or the overview, as can be seen in Hallström's chapter in the discussion of atomistic versus systemic conceptions. Either way, the four points could involve both the reduction of detail, or the mapping out of complexity, as suggested by the Vries.

Another important aspect of teaching and learning about technological systems that can be derived from the research presented in this book is the importance of a relevant *progression* of student learning. Hallström, Klasander and Zetterqvist point to the importance of, and start the development of, such a progression, but it is still nascent concerning the learning of technological systems. However, a classical progression line, as suggested above, is to have younger students learn more basic knowledge and thereafter increase complexity of systems content with the students' age. For example, as suggested by Barak, different aspects of feedback can be learned at different levels, from more qualitative aspects for younger students to more mathematical knowledge for older students. Norström also points to the importance of relevant examples of technological systems for teaching. It is crucial to pick those that are most clear and with the most potential for learning, so in terms of a progression examples should include fewer components and less complexity for younger students and could have more components and be more complex as the students age (cf. Engström & Svensson's chapter). However, it is important here to point to the concept of systems horizon, which is the border between what can be designated as a technological artefact and what is considered to be a technological system. Even younger students need to be introduced to systems, so one task for students of this age could be to identify both artefacts and simple systems, and what the differences are (see de Vries' chapter, and Hallström & Klasander, 2020).

A third aspect of teaching and learning about technological systems pointed out in chapters by, for instance, Barak, Norström, Svensson, Mioduser, and Engström and Svensson, is the usefulness of *modelling*, both with models derived from technology or engineering and educational models. Although employed in different ways, the Structure, Behaviour, Function (SBF) model features in several chapters, most prominently in Svensson's and Mioduser's, but one can also perhaps see the Freiburg educational model in Engström and Svensson's chapter as a variant with its focus on structural, behavioural, functional and contextual aspects of technological systems. The variations of the SBF model could thus help students explore intrinsic features of technological systems in relation to the systems' structure, behaviour, function—and context.

The different interpretations of the SBF model concerning what should be included in the structure, behaviour, function (and context) categories respectively in these two (or three) chapters should definitely be regarded as a strength because they point to the plasticity of the model. It is apparently effective by allowing for dynamic and nuanced interpretations of aspects or layers that are common to more than one technological system, thereby allowing for structured comparisons between systems. This, in turn, is important for students to be able to generalize knowledge about technological systems and learn about what are the most common characteristics of technological systems and how they differ from single artefacts. Such generalized knowledge could be called systems thinking, which is a way of approaching the study of technological (and other) systems that features prominently in chapters by Svensson, Engström and Svensson, and Chatoney and Gunther. It is also conceivable to view systems thinking as a model, as in the two former chapters.

Some of the chapters of the book, for instance, Hallström's and Dagan's, also report on teaching interventions which can be used as examples of how to go about teaching of detailed knowledge about technological systems. Other chapters such as Hallström and Kaijser's suggest and exemplify concepts with appropriate pedagogical and disciplinary affordance (Airey & Linder, 2017) for an optimal teaching of systems. In conclusion, by employing the appropriate *simplification*, *progression* and *modelling* when teaching, there can also be effective student learning about technological systems.

References

Airey, J., & Linder, C. (2017). Social semiotics in university physics education. In D. F. Treagust, R. Duit, & H. E. Fischer (Eds.), *Multiple representations in physics education* (pp. 95–122). Springer.

Arbesman, S. (2017). *Overcomplicated: Technology at the limits of comprehension*. Portfolio.

Bertalanffy, L. V. (1973). *General system theory: Foundations, development, applications* (Rev. ed.). Braziller.

Compton, V. J. (2019). Günter Ropohl: supporting a technological literacy for future citizenship. In J. R. Dakers, J. Hallström, & M. J. de Vries (Eds.), *Reflections on technology for educational practitioners: Philosophers of technology inspiring technology education* (pp. 37–54). Brill Sense.

Compton, V. J., & Compton, A. D. (2013). Teaching technological knowledge: Determining and supporting student learning of technological concepts. *International Journal of Technology and Design Education, 23*(3), 637–674.

de Vries, M. J., Fletcher, S., Kruse, S., Labudde, P., Lang, M., Mammes, I., Max, C., Münk, D., Nicholl, B., Strobel, J., & Winterbottom, M. (Eds.). (2020). *The impact of technology education: International insights*. Waxmann.

Ginns, I. S., Norton, S. J., & McRobbie, C. J. (2005). Adding value to the teaching and learning of design and technology. *International Journal of Technology and Design Education, 15*(1), 47–60.

Hallström, J. (2009). Systemteori och teknik. En introduktion till stora tekniska system. In Gyberg, P., & Hallström, J. (Eds.), *Världens gång—teknikens utveckling. Om samspelet mellan teknik, människa och samhälle* (pp. 99–117). Lund, Studentlitteratur.

Hallström, J., & Klasander, C. (2017). Visible parts, invisible whole: Swedish technology student teachers' conceptions about technological systems. *International Journal of Technology and Design Education, 27*(3), 387–405. https://doi.org/10.1007/s10798-016-9356-1

Hallström, J., & Klasander, C. (2020). Making the invisible visible: Pedagogies related to teaching and learning about technological systems. In P. J. Williams & D. Barlex (Eds.), *Pedagogy for technology education in secondary schools: Research informed perspectives for classroom teachers* (pp. 65–82). Springer.

Hmelo-Silver, C. E., & Azevedo, R. (2006). Understanding complex systems: Some core challenges. *The Journal of the Learning Sciences, 15*(1), 53–61.

Ho, F. M. (2019). Turning challenges into opportunities for promoting systems thinking through chemistry education. *Journal of Chemical Education, 96*(12), 2764–2776.

Hughes, T. P. (1983). *Networks of power: Electrification in Western society, 1880–1930*. Johns Hopkins University Press.

Hughes, T. P. (1987). The evolution of large technological systems. In W. E. Bijker, T. P. Hughes, T. J. Pinch (Eds.), *The social construction of technological systems: New directions in the sociology and history of technology*. The MIT Press.

Jacobson, M. J., & Wilensky, U. (2006). Complex systems in education: Scientific and educational importance and implications for the learning sciences. *The Journal of the Learning Sciences, 15*(1), 11–34.

Jones, A., Buntting, C., & de Vries, M. J. (2013). The developing field of technology education: A review to look forward. *International Journal of Technology and Design Education, 23*(2), 191–212.

Klasander, C. (2006). Innehållsaspekter av systemperspektiv i teknikundervisning. En studie av sex olika nationers styrdokument. Paper presented at the *Forskningskonferensen "Styrdokument och klassrumsverklighet i skolans teknikundervisning"*. http://www.ep.liu.se/ecp/017/004/ecp017004.pdf

Klasander, C. (2010). Talet om tekniska system. Förväntningar, traditioner och skolverkligheter. Linköpings universitet.

Koski, M.-I., & de Vries, M. J. (2013). An exploratory study on how primary pupils approach systems. *International Journal of Technology and Design Education, 23*(4), 835–848. https://doi.org/10.1007/s10798-013-9234-z

Kroes, P., Franssen, M., van de Poel, I., & Ottens, M. (2006). Treating socio-technical systems as engineering systems: Some conceptual problems. *Systems Research and Behavioral Science, 23*, 803–814. https://doi.org/10.1002/sres.703

Lind, J., Pelger, S., & Jakobsson, A. (2019). Students' ideas about technological systems interacting with human needs. *International Journal of Technology and Design Education, 29*, 263–282. https://doi.org/10.1007/s10798-018-9449-0

Mioduser, D., Venezky, R. L., & Gong, B. (1996). Students' perceptions and designs of simple control systems. *Computers in Human Behavior, 12*(3), 363–388.

Pearson, G., & Young, A. T. (Eds.). (2002). *Technically speaking: Why all Americans need to know more about technology*. National Academies Press.

Plate, R. (2010). Assessing individuals' understanding of nonlinear causal structures in complex systems. *Computers in Human Behavior, 26*(1), 19–33.

Riess, W., Schuler, S., & Hoersch, C. (2015). Wie laesst sich systemisches Denken vermitteln und foerdern? Theoretische Grundlagen und praktische Umsetzung am Beispiel eines Seminars fuer Lehramtsstudierende [How can systems thinking be taught and encouraged? Theoretical foundation and practical application in a course for teacher students]. *Geographie Aktuell Und Schule, 37*, 16–29.

Ropohl, G. (1997). Knowledge types in technology. *International Journal of Technology and Design Education, 7*, 65–72.

Ropohl, G. (2009). *Allgemeine Technologie. Eine Systemtheorie der Technik* (3rd ed.). Universitätsverlag Karlsruhe.

Rosenkränzer, F., Kramer, T., Hörsch, C., Schuler, S., & Riess, W. (2016). Promoting student teachers' content related knowledge in teaching systems thinking: Measuring effects of an intervention through evaluating a videotaped lesson. *Higher Education Studies, 6*(4), 156–169.

Schooner, P., Klasander, C., & Hallström, J. (2018a). Swedish technology teachers' views on assessing student understandings of technological systems. *International Journal of Technology and Design Education, 28*(1), 169–188. https://doi.org/10.1007/s10798-016-9383-y

Schooner, P., Klasander, C., & Hallström, J. (2018b). Teknik, systemgräns och människa: Tekniklärares uppfattningar om vad tekniska system är. *NorDiNa. Nordic Studies in Science Education, 14*(4).

Slangen, L., van Keulen, H., & Gravemeijer, K. (2011). What pupils can learn from working with robotic direct manipulation environments. *International Journal of Technology and Design Education, 21*(4), 449–469.

Svensson, M. (2011). *Att urskilja tekniska system. Didaktiska dimensioner i grundskolan*. Linköpings universitet.

Svensson, M. (2018). Learning about systems. In M. J. de Vries (Ed.), *Handbook of technology education* (pp. 447–462). Springer International Publishing.

Svensson, M., & Klasander, C. (2012). Teacher's professional growth in planning and teaching technological systems. Paper presented at the *Technology Education Research Conference, Surfers Paradise*, Australia.

Verkerk, M. J., Hoogland, J., Van der Stoep, J., & de Vries, M. J. (Eds.). (2016). *Philosophy of technology: An introduction for technology and business students*. Routledge.

Vermaas, P., Kroes, P., van de Poel, I., Franssen, M., & Houkes, W. (2011). *A philosophy of technology: From technical artefacts to sociotechnical systems*. Morgan & Claypool Publishers.

Wiener, N. (1948). *Cybernetics: Or control and communication in the animal and the machine*. Hermann & Co., Editeurs.

Jonas Hallström is Ph.D. and Professor of Technology Education at the Department of Behavioural Sciences and Learning, Linköping University, Sweden, where he is also currently Deputy Head of Department. He is also Chair of the Swedish National Graduate School in Science and Technology Education Research (FontD). He presents regularly at international conferences and consults on technology education. He is one of the editorial board members of the Brill/Sense book series *International Technology Education Studies*. His research primarily concerns the historical emergence of technology as knowledge content in the school, the epistemology and subject philosophy of technology, various subject content (e.g. technological systems) as well as the attitudes to and knowledge of technology and technology education of students, student teachers and teachers. The research also relates to technology teaching in relation to, e.g. design, gender (girls and technology), authentic learning, models and modelling and STEM (science, technology, engineering, mathematics) education.